Parasitic Weeds: Biology and Control

Parasitic Weeds: Biology and Control

Editors

Evgenia Dor
Yaakov Goldwasser

MDPI • Basel • Beijing • Wuhan • Barcelona • Belgrade • Manchester • Tokyo • Cluj • Tianjin

Editors

Evgenia Dor
Newe Ya'ar Research Center
Israel

Yaakov Goldwasser
The Hebrew University of Jerusalem
Israel

Editorial Office
MDPI
St. Alban-Anlage 66
4052 Basel, Switzerland

This is a reprint of articles from the Special Issue published online in the open access journal *Plants* (ISSN 2223-7747) (available at: https://www.mdpi.com/journal/plants/special_issues/parasitic_weeds).

For citation purposes, cite each article independently as indicated on the article page online and as indicated below:

LastName, A.A.; LastName, B.B.; LastName, C.C. Article Title. *Journal Name* **Year**, *Volume Number*, Page Range.

ISBN 978-3-0365-5289-7 (Hbk)
ISBN 978-3-0365-5290-3 (PDF)

Cover image courtesy of Yaakov Goldwasser

Contents

Preface to "Parasitic Weeds: Biology and Control"

This Special Issue is focused on both research articles and reviews providing new insight into the field of parasitic plants, including topics such as the biology and life cycle of parasitic plants, the physiology of parasitism, genetics, evolution, population dynamics, host–parasite interaction, etc.

Evgenia Dor and Yaakov Goldwasser
Editors

Editorial

"Parasitic Weeds: Biology and Control" Special Issue Editors Summary

Evgenia Dor [1],* and Yaakov Goldwasser [2],*

[1] Institute of Plant Protection, Newe Ya'ar Research Center, Agricultural Research Organization, P.O. Box 1021, Ramat Yishay 30095, Israel
[2] The Robert H. Smith Institute of Plant Sciences and Genetics in Agriculture, Faculty of Agriculture, Food and Environment, The Hebrew University of Jerusalem, P.O. Box 12, Rehovot 76100, Israel
* Correspondence: evgeniad@volcani.agri.gov.il (E.D.); yaakov.goldwasser@mail.huji.ac.il (Y.G.)

Citation: Dor, E.; Goldwasser, Y. "Parasitic Weeds: Biology and Control" Special Issue Editors Summary. *Plants* **2022**, *11*, 1891. https://doi.org/10.3390/plants11141891

Received: 27 June 2022
Accepted: 19 July 2022
Published: 21 July 2022

Publisher's Note: MDPI stays neutral with regard to jurisdictional claims in published maps and institutional affiliations.

We are happy to summarize this important Special Issue (SI) of MDPI *Plants*—"Parasitic Weeds: Biology and Control".

Parasitic plants are scientifically interesting and agriculturally important weeds which are spreading worldwide with limited control means. Plant parasitism is a case of extreme plant-to-plant interactions when parasitic plants connect directly to the vasculature system of a host plant, extracting water and nutrients from them, and assimilates [1]. During the evolution from non-parasitic origin, parasitic plants have developed many specific functions, such as host detection, host attachment, host exploitation, and host defense suppression. The world of parasitic plants includes about 20 families, with a wide trophic spectrum from facultative hemiparasites, which are able also to perform photosynthesis and therefore may survive without a connection to the host, to obligatory holoparasites, which have no photosynthetic abilities [2]. Some parasitic species are noxious weeds and damage major agricultural crops, causing heavy economical losses worldwide [3,4].

The parasitic lifestyle in plants has always been the subject of curiosity of scientists, but during the last decade our understanding of parasitic plant—host interaction has greatly evolved due to rapid advances in molecular and genomic tools, especially such as high throughput DNA sequencing, transcriptomics and metabolomics. The latest findings take the science of parasitic plants to a higher level and open up new horizons in parasitic plant management. The discovery of a novel family of phytohormones, the strigolactones [5,6], and their involvement in host detection and evolution of parasitic plants [7,8], the detection of the information exchange between host and parasite [9]; and the elucidation of the host defense mechanisms suppression by the parasites [10] have led to a deeper understanding of physiological processes in host-parasite interaction. In the light of recent achievements, the re-evaluation of control management, including crop breeding, and molecular genetics is on the agenda.

Finally, 11 papers were collected for the SI; of them, five original research papers present new strategies in parasitic weeds management [11–15], two focused on crop resistance to parasitic plants [16,17] and three provide new insights on plant—parasite interaction [18–20]. One Opinion Paper provides a personal view of the present status of parasitic weed problems and their control written by Chris Parker, who dedicated his entire long scientific career to parasitic plants and their management [21].

We thank all the authors for their valuable articles. We are especially proud of the active participation of young scientists in our Special Issue. Four articles were submitted for participation in the student paper competition [11,14,15,17]. The winner of the competition was the Ph.D. student Dana Sisou from the Department of Phytopathology and Weed Research, ARO, Newe Ya'ar Research Center. The title of her paper was based on her Ph.D. thesis: Biological and transcriptomic characterization of pre-haustorial resistance to sunflower broomrape (*Orobanche cumana* W.) in sunflowers (*Helianthus annuus*) [17].

We would particularly like to acknowledge Chris Parker's paper contribution, the parasitic plant elder on one end, and the inclusion of student papers on the other end—we feel that we have successfully encouraged both and those in between to contribute to this important field.

Finally, we offer special thanks to Mrs. Sumi Sun, the MDPI *Plants* Special Issue coordinator for her patience, good work, and assistance with the editing and publication processes.

Funding: This research received no external funding.

Conflicts of Interest: The author declares no conflict of interest.

References

1. Heide-Jørgensen, H.S. Introduction: The parasitic syndrome in higher plants. In *Parasitic Orobanchaceae. Parasitic Mechanisms and Control Strategies*; Joel, D.M., Gressel, J., Musselman, L.J., Eds.; Springer: Berlin/Heidelberg, Germany, 2013; pp. 1–18.
2. Joel, D.M. The Orobanchaceaea and their parasitic mechanism. In *Parasitic Orobanchaceae. Parasitic Mechanisms and Control Strategies*; Joel, D.M., Gressel, J., Musselman, L.J., Eds.; Springer: Berlin/Heidelberg, Germany, 2013; pp. 21–23.
3. Gressel, J.; Joel, D.M. Weedy Orobanchaceae: The problem. In *Parasitic Orobanchaceae. Parasitic Mechanisms and Control Strategies*; Joel, D.M., Gressel, J., Musselman, L.J., Eds.; Springer: Berlin/Heidelberg, Germany, 2013; pp. 309–312.
4. Parker, C.; Riches, C.R. *Parasitic Weeds of the World: Biology and Control*; CAB International: Wallingford, UK, 1993; p. 332.
5. Xie, X.; Yoneyama, K.; Yoneyama, K. The strigolactone story. *Annu. Rev. Phytopathol.* **2010**, *48*, 93–117. [CrossRef] [PubMed]
6. Gomez-Roldan, V.; Fermas, S.; Brewer, P.B.; Puech-Pagès, V.; Dun, E.A.; Pillot, J.P.; Letisse, F.; Matusova, R.; Danoun, S.; Portais, J.C.; et al. Strigolactone inhibition of shoot branching. *Nature* **2008**, *455*, 189–194. [CrossRef] [PubMed]
7. Fernández-Aparicio, M.; Yoneyama, K.; Rubiales, D. The role of strigolactones in host specificity of *Orobanche* and *Phelipanche* seed germination. *Seed Sci. Res.* **2011**, *21*, 55–61. [CrossRef]
8. Thorogood, C.; Hiscock, S. Specific developmental pathways underlie host specificity in the parasitic plant *Orobanche*. *Plant Signal Behav.* **2010**, *5*, 275–277. [CrossRef]
9. Koltai, H. Cellular events of strigolactone signalling and their crosstalk with auxin in roots. *J. Exp. Bot.* **2015**, *66*, 4855–4861. [CrossRef]
10. Brun, G.; Thoiron, S.; Braem, L.; Pouvreau, J.B.; Montiel, G.; Lechat, M.M.; Simier, P.; Gevaert, K.; Goormachtig, S.; Delavault, P. CYP707As are effectors of karrikin and strigolactone signalling pathways in Arabidopsis thaliana and parasitic plants. *Plant Cell Environ.* **2019**, *42*, 2612–2626. [CrossRef]
11. Fatino, M.J.; Hansosn, B.D. Evaluating branched broomrape (*Phelipanche ramosa*) management strategies in California processing tomato (*Solanum lycopersicum*). *Plants* **2022**, *11*, 438. [CrossRef] [PubMed]
12. Goldwasser, Y.; Rabinovitz, O.; Gerstl, O.; Nasser, A.; Paporisch, A.; Kuzikaro, H.; Sibony, M.; Rubin, B. Imazapic herbigation for Egyptian broomrape (*Phelipanche aegyptiaca*) control in processing tomatoes—laboratory and greenhouse studies. *Plants* **2021**, *10*, 1182. [CrossRef] [PubMed]
13. Jamil, M.; Wang, J.Y.; Yonli, D.; Patil, R.H.; Riyazaddin, M.; Gangashetty, P.; Berqdar, L.; Chen, G.-T.E.; Traore, H.; Margueritte, O.; et al. A new formulation for strigolactone suicidal germination agents, towards successful *Striga* management. *Plants* **2022**, *11*, 808. [CrossRef] [PubMed]
14. Miura, H.; Ochi, R.; Nishiwaki, H.; Yamauchi, S.; Xie, X.; Nakamura, H.; Yoneyama, K.; Yoneyama, K. Germination Stimulant Activity of Isothiocyanates on *Phelipanche* spp. *Plants* **2022**, *11*, 606. [CrossRef] [PubMed]
15. Wallach, A.; Achdari, G.; Eizenberg, H. Good news for cabbageheads: Controlling *Phelipanche aegyptiaca* infestation under hydroponic and field conditions. *Plants* **2022**, *11*, 1107. [CrossRef] [PubMed]
16. Galil, S.; Hershenhorn, J.; Smirnov, E.; Yoneyama, K.; Xie, X.; Amir-Segev, O.; Bellalou, A.; Dor, E. Characterization of a chickpea mutant resistant to *Phelipanche aegyptiaca* Pers. and *Orobanche crenata* Forsk. *Plants* **2021**, *10*, 2552. [CrossRef] [PubMed]
17. Sisou, D.; Tadmor, Y.; Plakhine, D.; Ziadna, H.; Hübner, S.; Eizenberg, H. Biological and transcriptomic characterization of pre-haustorial resistance to sunflower broomrape (*Orobanche cumana* W.) in sunflowers (*Helianthus annuus*). *Plants* **2021**, *10*, 1810. [CrossRef] [PubMed]
18. Chen, B.J.W.; Xu, J.; Wang, X. Trophic transfer without biomagnification of cadmium in a soybean-dodder parasitic system. *Plants* **2021**, *10*, 2690. [CrossRef] [PubMed]
19. Kumar, K.; Amir, R. The effect of a host on the primary metabolic profiling of *Cuscuta campestris*' main organs, haustoria, stem and flower. *Plants* **2021**, *10*, 2098. [CrossRef] [PubMed]
20. Park, S.-O.; Shimizu, K.; Brown, J.; Aoki, K.; Westwood, J.H. Mobile host mRNAs are translated to protein in the associated parasitic plant *Cuscuta campestris*. *Plants* **2022**, *11*, 93. [CrossRef] [PubMed]
21. Parker, C. A personal history in parasitic weeds and their control. *Plants* **2021**, *10*, 2249. [CrossRef] [PubMed]

 *plants*

Article

Imazapic Herbigation for Egyptian Broomrape (*Phelipanche aegyptiaca*) Control in Processing Tomatoes—Laboratory and Greenhouse Studies

Yaakov Goldwasser [1,*], Onn Rabinovitz [2], Zev Gerstl [3], Ahmed Nasser [3], Amit Paporisch [1], Hadar Kuzikaro [1], Moshe Sibony [1] and Baruch Rubin [1]

[1] R.H. Smith Institute of Plant Science & Genetics in Agriculture, R.H. Smith Faculty of Agriculture, Food and Environment, The Hebrew University of Jerusalem, P.O. Box 12, Rehovot 76100, Israel; kapitush@gmail.com (A.P.); hadar_koz@walla.co.il (H.K.); moshe.sibony@mail.huji.ac.il (M.S.); rubin@mail.huji.ac.il (B.R.)

[2] The Agricultural Extension Service, The Ministry of Agriculture, P.O. Box 28, Bet Dagan 50200, Israel; onnrab@gmail.com

[3] Institute of Soil, Water and Environmental Sciences, Volcani Center, ARO, P.O. Box 6, Bet Dagan 50250, Israel; zgerstl@volcani.agri.gov.il (Z.G.); nasser@volcani.agri.gov.il (A.N.)

* Correspondence: yaakov.goldwasser@mail.huji.ac.il

Citation: Goldwasser, Y.; Rabinovitz, O.; Gerstl, Z.; Nasser, A.; Paporisch, A.; Kuzikaro, H.; Sibony, M.; Rubin, B. Imazapic Herbigation for Egyptian Broomrape (*Phelipanche aegyptiaca*) Control in Processing Tomatoes—Laboratory and Greenhouse Studies. *Plants* **2021**, *10*, 1182. https://doi.org/10.3390/plants10061182

Academic Editor: Per Kudsk

Received: 9 May 2021
Accepted: 8 June 2021
Published: 10 June 2021

Publisher's Note: MDPI stays neutral with regard to jurisdictional claims in published maps and institutional affiliations.

Abstract: Parasitic plants belonging to the Orobanchaceae family include species that cause heavy damage to crops in Mediterranean climate regions. *Phelipanche aegyptiaca* is the most common of the Orobanchaceae species in Israel inflicting heavy damage to a wide range of broadleaf crops, including processing tomatoes. *P. aegyptiaca* is extremely difficult to control due to its minute and vast number of seeds and its underground association with host plant roots. The highly efficient attachment of the parasite haustoria into the host phloem and xylem enables the diversion of water, assimilates and minerals from the host into the parasite. Drip irrigation is the most common method of irrigation in processing tomatoes in Israel, but the delivery of herbicides via drip irrigation systems (herbigation) has not been thoroughly studied. The aim of these studies was to test, under laboratory and greenhouse conditions, the factors involved in the behavior of soil-herbigated imazapic, and the consequential influence of imazapic on *P. aegyptiaca* and tomato plants. Dose-response Petri dish studies showed that imazapic does not impede *P. aegyptiaca* seed germination and non-attached seedlings, even at the high rate of 5000 ppb. Imazapic applied to tomato roots inoculated with *P. aegyptiaca* seeds in a PE bag system revealed that the parasite is killed only after its attachment to the tomato roots, at concentrations as low as 2.5 ppb. Imazapic sorption curves and calculated Kd and Koc values indicated that the herbicide Kd is similar in all soils excluding a two-fold higher coefficient in the Gadash farm soil, while the Koc was similar in all soils except the Eden farm soil, in which it was more than twofold lower. In greenhouse studies, control of *P. aegyptiaca* was achieved at >2.5 ppb imazapic, but adequate control requires repeated applications due to the 7-day half-life ($t_{1/2}$) of the herbicide in the soil. Tracking of imazapic in soil and tomato roots revealed that the herbicide accumulates in the tomato host plant roots, but its movement to newly formed roots is limited. The data obtained in the laboratory and greenhouse studies provide invaluable knowledge for devising field imazapic application strategies via drip irrigation systems for efficient and selective broomrape control.

Keywords: chemigation; drip irrigation; Egyptian broomrape; herbicide; imazapic; parasitic plants; tomato; weed control

1. Introduction

Parasitic plants account for approximately one percent of angiosperm species and are present in 275 genera belonging to 28 botanical families [1]. Some of the parasitic plants are important agricultural weeds, infest a varied range of crops worldwide, and pose a threat to the food security of many communities [2–4]. The parasitic plant family Oroban-chaceae includes 270 species in 20 genera of root holoparasites of which the two Oroban-chaceae genera, *Orobanche* and *Phelipanche* (common name broomrape), include more than 100 species, a few of which are major agricultural weeds in regions with Mediterranean climates. *Phelipanche aegyptiaca* is a major obstacle to the production of many broadleaf crops, most of them belonging to the Solanaceae, Fabaceae, Cruciferae and Umbelliferae plant families [5,6]. Outbreaks of *P. aegyptiaca* infestations occur frequently outside of the Mediterranean region and have been reported in Africa, Asia, Europe and recently in North America [7,8].

Broomrape species are extremely difficult to control during the cultivation of agricul-tural crops due to the intimate hidden underground association of the parasite haustoria with the roots of the host plant. The parasite acts as a strong metabolic sink that sucks assimilates, water and minerals from the host plant's vascular system. Numerous meth-ods have been attempted for field broomrape control including hand weeding, chemical herbicides, bioherbicides and biological agents, and naturally and genetically engineered crop resistance [5,6,9–11], most of them resulting in limited success under field-application conditions. The use of chemical control as reported in this study requires a highly effective and selective herbicide to control the parasite without harming the crop.

Imazapic belongs to the imidazolinone herbicide group, a subgroup of the aceto-lactate synthase enzyme inhibiting herbicides (ALS), systemic herbicides that hinder the production of branched-chain aliphatic amino acids necessary for protein synthesis and cell growth [12]. Some of these herbicides have been shown to kill broomrape directly in the soil and systemically via the host plant as the parasite acts as a strong sink readily absorbing the herbicides from the host plant [13–15]. Imazapic is a weak acid (pKa $\cong$ 3.6) that behaves as an anion in most soils [16]. The herbicide is highly soluble (2200 mg/L @25 °C) and its reported half-life in the soil varies from 31 to 233 days, depending upon soil characteristics and environmental conditions [17]. Imazapic is weakly adsorbed in high pH soil, while adsorption increases as pH decreases and as clay and organic matter content increase [17]. The herbicide is not volatile and there is little lateral movement of the herbicide in soil. The half-life of the herbicide on soils due to photolysis is 120 days but in aqueous solutions, imazapic is rapidly broken down by photolysis with a half-life of just one or two days [12,17,18].

The most common irrigation system in Israeli tomato production is surface located drip irrigation in which water and nutrients are directly and accurately applied to the plant root system. However, the distribution of herbicides in the soil under drip irrigation will vary under different soil conditions and irrigation regimes. The hypothesis proposed by us is that the parasitic plant *P. aegyptiaca* can be controlled efficiently and selectively via chemigation of imazapic through drip irrigation (herbigation) which enables elegant and precise delivery of the herbicide to the parasite.

The aim of this study was to test, under laboratory and greenhouse conditions, the factors involved in the behavior of imazapic in soil, and the consequential phytotoxicity of the herbicide to *P. aegyptiaca* and tomato plants, enabling the devising and refinement of a management tool that will effectively and selectively control *P. aegyptiaca* in processing tomato under field conditions.

2. Materials and Methods

2.1. Phelipanche Aegyptiaca Seed

The source of *P. aegyptiaca* seeds used in the laboratory and greenhouse studies was from a 2002 heavily infested chickpea (*Cicer arietinum*) field in Gesher Haziv, located in the Western Galilee of Israel (N33°04′07″, E35°11′00″No). Capsules were removed from mature dry broomrape inflorescences, air-dried, threshed, cleaned and stored in a plastic container at 4 °C until use.

For Petri dish and laboratory studies, seeds were disinfected by soaking in 70% ethanol for one minute and then in sodium hypochlorite 1% + 'Tween 20' 0.01% (by volume) for ten minutes. The seeds were then washed five times with sterilized water.

2.2. Imazapic Herbicide

The commercial imazapic product 'Cadre' 240 g/L produced by BASF Germany was used in all experiments. This herbicide is a selective systemic ALS inhibitor belonging to the imidazolinone group, registered and sold in Israel by Luxembourg Industries Ltd., Tel-Aviv 6812509, Israel.

2.3. LC-MS/MS Imazapic Analysis

LC-MS/MS chemical analysis of imazapic in soil and tomato roots was conducted according to protocols developed by us during this study. Imazapic analysis was carried out on a Sciex AB 3200 Qtrap LCMS using ESI. Ten μL were injected onto a Kinetex® 5 μm C18 100 Å, 150 × 4.60 mm Phenomenex column. The mobile phase consisted of 30% eluent A and 70% eluent B: eluent A-0.2% acetic acid in water and eluent B- a 1:1 mixture of methanol:acetonitrile. The flow rate was 0.8 mL/min. Imazapic was identified by MRM with the following transitions: 276.1 → 231.1 and 276.1 → 86. The described analysis enables the detection of imazapic concentrations as low as 1 ppb.

2.4. Soil Samples Analysis

Imazapic was extracted from soil according to the following procedure: Ten ml of a water:methanol solution (70:30) were added to 3–5 g of soil. The sample was vortexed for 30 s and then shaken overnight on a reciprocal shaker. The samples were then centrifuged at an RCF of 2500× *g* and about 2 mL of the clear supernatant were transferred to an LCMS vial after filtration through a 45 μ spiral filter.

2.5. Tomato Root Samples Analysis

Tomato root samples were prepared for LC-MS/MS analysis in the following manner: plant material was dried in a lyophilizer and then ground with a mortar and pestle. A 0.1 g sample was taken and extracted with 5 mL of a 1:1 water:acetonitrile mixture and 2 mL of hexane. The sample was vortexed for 30 sec and then shaken overnight on a reciprocal shaker. After centrifugation, about 2 mL of the aqueous phase were transferred to an LCMS vial after filtration through a 45 μ spiral filter. LC-MS/MS analysis was carried out as described above.

2.6. Dose-Response of P. aegyptiaca Seed and Seedlings to Imazapic in Petri Dish

To test the susceptibility of *P. aegyptiaca* to imazapic in the very early parasite developmental stage we conducted dose-response experiments in which we applied imazapic at increasing concentrations to *P. aegyptiaca* seeds and seedlings.

The procedure for the seed germination experiment was as follows: following seed disinfection, ~100 *P. aegyptiaca* seeds were sprinkled on five cm diameter glass-microfibre filter discs, placed in Petri dishes and moistened with 700 μL sterilized water. The Petri dishes were then sealed with Parafilm, covered with aluminum foil, and placed in a 25 °C growth chamber for a pre-conditioning period. After seven days the Petri dishes were opened and treated with 0–5000 ppb imazapic solutions containing 500 μL of the synthetic strigol analog germination stimulant GR24. The Petri dishes were then re-sealed with parafilm, wrapped with aluminum foil, and placed back in the growth chamber. After an additional 14 days, *P. aegyptiaca* seed germination in each Petri dish was determined under a stereoscopic microscope: seeds in which the hypocotyl was longer than the seed length were considered germinating seeds.

The procedure for the seedling control experiment was as follows: seeds were sprinkled and treated as in the seed germination experiment but after one week only the synthetic germination stimulant GR24 was added (no imazapic). One week later, following seed germination, Petri dishes were opened and imazapic was added to the Petri dishes at the same concentrations as in the seed germination experiment. The Petri dishes were then resealed, covered and placed in the growth chamber. One week later seedling vitality was determined under the stereoscopic microscope.

All treatments and controls were replicated 5 times. Treatments were accompanied by two controls: 1. The synthetic stimulant GR24 was added to the Petri dish with no imazapic to test the potential germination of the seeds. 2. Five hundred µL of sterilized water only was added after preconditioning to test for spontaneous germination. All procedures were conducted under sterile conditions in a laminar flow hood to prevent contamination of seeds and seedlings.

2.7. Effect of Imazapic on P. aegyptiaca Parasitizing Tomato Roots in Polyethylene Bag Studies

In addition to the Petri dish experiments in which we tested the effect of imazapic on *P. aegyptiaca* seeds, imazapic was also applied to *P. aegyptiaca* attachments and tubercles after their attachment to tomato roots in a polyethylene bag system (Goldwasser et al., 1997). This system allows us to non-destructively monitor the development of the parasite on host plant roots and the effect of the herbicide on both the parasite and the host plant. After seed disinfection (as described above), 10 mg of *P. aegyptiaca* seeds were sprinkled onto 14 by 12 cm glass-microfibre sheets (GFA paper), previously moistened with 5 mL sterile water. One 4-week-old tomato seedling was mounted on the top of each GFA sheet. Sheets were then inserted into a clear polyethylene bag (25 by 18 cm), and 100 mL of sterilized Hoagland nutrient solution was added to each PE bag (Figure 1). Polyethylene bags were hung upright in a black box so that plant roots were in the dark and their shoots projected into the air and light above the box. The box was placed in a 25 °C growth chamber and the nutrient solution was replenished in the bags as needed. Thirty-six days after planting (DAP), PE bags were emptied of the nutrient solution and imazapic was injected by a syringe to each bag mixed in 20 mL Hoagland nutrient solution at 2.5, 5 and 10 ppb, with 10 replications per treatment. The PE bags were placed in the 25 °C growth chamber and weekly monitored and recorded for broomrape and host root and foliage development.

Figure 1. The PE bag system for studying the effect of imazapic on *P. aegyptiaca* parasitism on tomato roots shown on the day of imazapic application 36 DAP.

2.8. Imazapic Sorption in Four Soils

To elucidate the differential efficacy of imazapic for *P. aegyptiaca* control in different soils, sorption of imazapic was studied in four soils taken from the following field sites: Ein Harod (Jezreel Valley), Eden Farm (Bet Shean), Gadash Farm (Upper Galilee) and Bet Dagan (Central Israel). Properties of the soils are presented in Table 1. Sorption was determined by the batch method: Ten ml of imazapic solution was added to 5 g of soil and shaken for 18 h on a reciprocal shaker. After centrifugation, a portion of the clear supernatant was analyzed by LC-MS/MS as described above. The initial concentrations of imazapic were 0 to 400 µg/L for the Eden Farm and Gadash Farm soils and 0 to 700 µg/L for the Ein Harod and Hamra soils. The experiment was run in duplicate and sorption was calculated based on the change in imazapic concentration in the supernatant.

Table 1. Properties of the soils in the imazapic sorption studies.

Soil Source	pH	Clay (%)	Silt (%)	Sand (%)	OC (%)
Eden Farm	7.4	47	27	26	1.2
Gadash Farm	7.5	59	31	10	0.94
Ein Harod	8.0	57	29	14	0.41
Bet Dagan	7.6	10	3	87	0.40

OC—Soil organic carbon.

2.9. Imazapic Dose-Response of Tomato and P. aegyptiaca in Pots

Following the imazapic dose-response of *P. aegyptiaca* seeds and seedlings in the Petri dish experiments and after attachment to tomato roots in the PE bag system, we tested the imazapic dose-response of the parasite and the host in pots in the greenhouse. The experiments were conducted in 3 L pots (19 cm diameter) filled with soil taken from the Upper Galilee Gadash Experimental Farm. *P. aegyptiaca* seeds were mixed in the soil with a cement mixer at a rate of 10 mg seeds L^{-1} soil. In each pot we planted a one-month-old tomato var. M-82 plug seedling. Imazapic was single and double drench-applied in 100 mL of water at different rates and timing of application, described in Table 2. Each treatment was replicated five times: five pots with one tomato plant each. Weekly assessment of tomato plants and broomrape inflorescences counts were performed and recorded. The experiment was terminated 71 days after tomato planting following a final count of broomrape inflorescences in each pot, cutting of each tomato plant at soil level and weighing fruit yield and foliage fresh weight for each plant.

Table 2. Imazapic treatments in the imazapic dose-response pot experiment. DAP = days after planting.

Treatment	Imazapic Application (ppb)	
	27 DAP	**51 DAP**
Control		
1	1.0	
2	2.5	
3	2.5	2.5
4	5.0	
5	5.0	5.0
6	7.5	
7	10.0	

2.10. Tracking Imazapic Concentration in Soil and Tomato Roots

The experiment was conducted in pots in the greenhouse as described for the dose-response experiment. Imazapic was applied 32 days after planting at 10 ppb as a single application in 100 mL water. Soil and root samples were collected from each pot 1, 3 and 7 DAA of imazapic. Soil samples and tomato root samples were prepared and analyzed for imazapic presence by LC-MS/MS as previously described.

2.11. Statistical Analysis

The data of each experiment was analyzed using JMP Pro® statistical software, Version 14, SAS Institute Inc., Cary, NC, USA 1989–2019.

Comparison of the means of each treatment in all tables and graphs was conducted using the standard error of the mean of each treatment or by statistical analysis using the Tukey Kramer HSD test, $\alpha = 0.05$ or the Student's t LSMeans Differences test, $\alpha = 0.05$.

The sorption isotherms for imazapic in different soils were determined by calculating the Kd value (L/kg), which describes the equilibrium between the herbicide concentration in the soil (mg/kg) relative to the water concentration (mg/L).

Statistical analysis of the Kd values of the different soils was determined by the Tukey Kramer HSD test, $p = 0.01$.

3. Results

3.1. Dose-Response of P. aegyptiaca Seed and Seedlings to Imazapic in Petri Dish

P. aegyptiaca seed germination was not affected by imazapic even at the extremely high concentration of 5000 ppb. At this herbicide dose, the parasite seeds germinated at a rate of 97–98% of the non-imazapic treated control. One-week-old *P. aegyptiaca* seedlings (with no association to the roots of a host plant) were not affected by any of the imazapic treatments as well (Table 3).

Table 3. Dose-response of *P. aegyptiaca* seed germination and seedling vigor to imazapic. Seed germination in the no-herbicide control treatment in which only the GR24 stimulant was added was 79.4%. No spontaneous seed germination was recorded in the no-GR24 no-imazapic control treatment. $\pm$ = Standard error.

Imazapic Concentration (ppb)	Seed Germination (% of Control)	Seedling Vigor (% of Control)
0	100.0 ± 0.01	100.0 ± 0.02
500	97.90 ± 0.04	94.98 ± 0.02
1000	96.54 ± 0.02	96.45 ± 0.03
5000	97.28 ± 0.01	98.45 ± 0.04

3.2. Effect of Imazapic on P. aegyptiaca Parasitizing Tomato Roots in Polyethylene Bag Studies

P. aegyptiaca seed germination and unattached parasite seedlings were not affected by the application of imazapic, a finding in agreement with the Petri dish studies. However, once the parasite was attached to the tomato host root it became susceptible to the herbicide application at both 2.5 and 5 ppb concentrations, as determined by the senescence of the tubercle and its subsequent detachment from the tomato host root (Figure 2).

3.3. Imazapic Sorption to Soils

The sorption isotherms for imazapic in the studied soils are presented in Figure 3. All isotherms were linear. The sorption coefficients Kd and the OC normalized sorption coefficients (Koc) for imazapic are presented in Table 4.

The significantly higher Kd found for the Gadash Farm soil indicates that more herbicide will be adsorbed by this soil, resulting in lower efficacy of *P. aegyptiaca* control compared to the other tested soils, possibly requiring a higher dose of imazapic for efficient parasite control. The high Kd of the Gadash farm soil can be explained by its high clay and organic matter contents. The significantly lower Kd value found for the Ein Harod soil denotes that less herbicide will be adsorbed by the soil and more herbicide is available in the soil solution, thus lower imazapic doses may be needed to control the pest in this soil compared to the other tested soils.

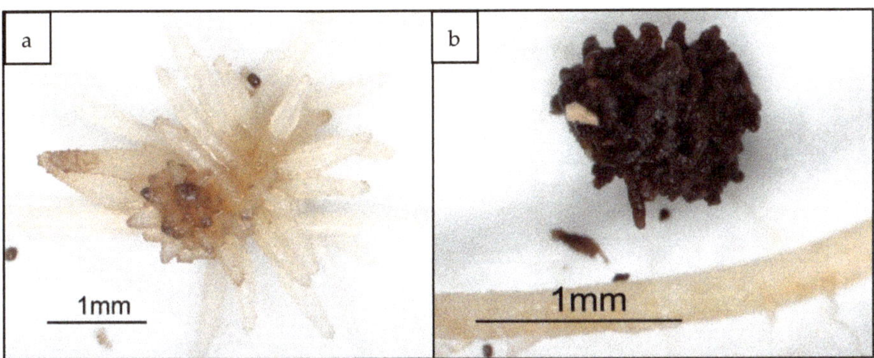

Figure 2. The effect of imazapic on *P. aegyptiaca* tubercles attached to tomato root in the PE bag system. (**a**)—Untreated control. (**b**)—Treated with 5 ppb imazapic.

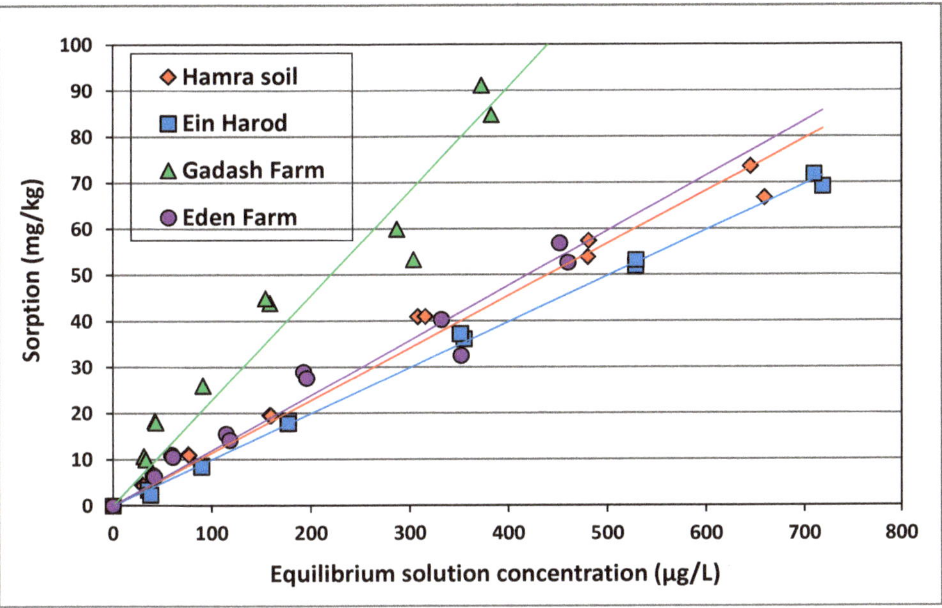

Figure 3. Imazapic sorption in three soils in which field studies were performed and in Hamra soil as a reference soil. All isotherms were linear.

Table 4. The calculated sorption coefficients. Kd, and the organic carbon (OC) normalized sorption coefficients (Koc) for imazapic in the four tested soils. Letters following the Kd values represent statistical differences according to the Tukey HSD test, $p = 0.01$.

Soil	Kd (L/kg)	Koc	r^2
Gadash Farm	0.23 a	24.4	0.938
Eden Farm	0.13 b	10.8	0.957
Hamra	0.11 bc	27.5	0.983
Ein Harod	0.10 c	26.8	0.998

3.4. Imazapic Dose-Response of Tomato and P. aegyptiaca in Pots

In the imazapic dose-response pot experiment, *P. aegyptiaca* emergence started in the 0, 2.5, 2.5 + 2.5 and 5.5 ppb treatments a few days after the first imazapic treatment (Figure 4). The number of inflorescences gradually increased in these treatments, reaching 4.7 to 5.7 inflorescences per pot 28 DAA of imazapic. In the 7.5 and 10 ppb treatments, *P. aegyptiaca* inflorescences emergence was delayed until 21 DAA and held to 1.6 and 0.83 inflorescences per pot 21 DAA, and reaching 3.17 and 2.17 at the end of the experiment, significantly lower than the infestation in the no-herbicide control treatment.

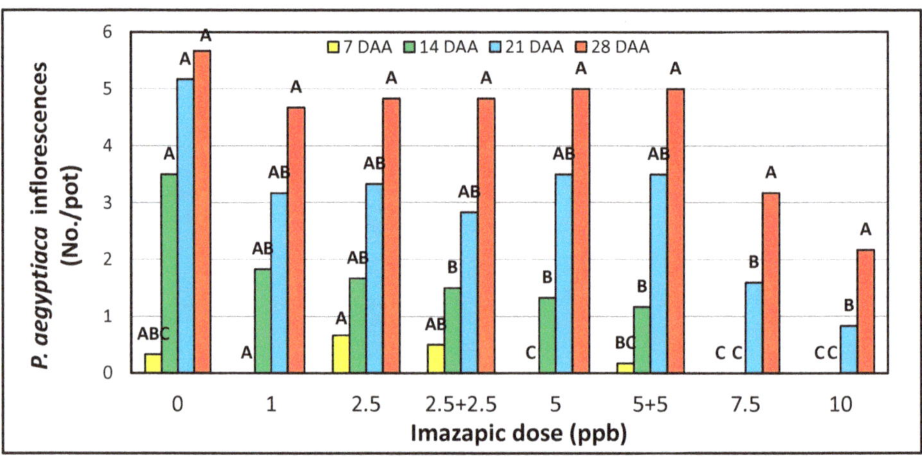

Figure 4. The accumulating number of emerging *P. aegyptiaca* inflorescences in the imazapic dose-response experiment in pots in the greenhouse. See Table 2 for the description of the different treatments. Letters above bars represent the statistical differences between the *P. aegyptiaca* inflorescences of each treatment analyzed by applying the Tukey-Kramer HSD test, α = 0.05. DAA-days after first imazapic application.

The effect of the higher doses of imazapic was more apparent on *P. aegyptiaca* growth than on the number of parasite inflorescences, as observed by the final inflorescences fresh weight (Figure 5). The inflorescences FW decreased from 21 g per pot in the non-treated control to 16.5–17.4 g per pot in the 1 and 2.5 ppb treatments, 12 to 11.9 g in the 2.5 + 2.5 and 5 ppb treatments, and 6.5 to 5.8 g in the 7.5 and 10 ppb treatments. All treatments reduced final *P. aegyptiaca* fresh weight compared to the non-treated control, but only the higher doses of 7.5 and 10 ppb caused a statistically significant reduction.

Though phytotoxic symptoms on tomato plants were observed for imazapic concentrations above 5 ppb, fruit yield was not statistically different between all treatments (data not presented).

3.5. Tracking Imazapic Concentration in Soil and the Tomato Plant

The soil and root imazapic analysis indicated that 1 DAA of imazapic the herbicide accumulated in the tomato roots, reaching concentrations of 263.6 ppb. Three DAA the accumulation in the roots continued, reaching a concentration 1.7-fold higher than at 1 DAA, while in the soil it dropped 2.9 fold. On day 7 the soil concentration diminished to 0.4 ppb and the root concentration dropped but remained high at 234 ppb (Table 5).

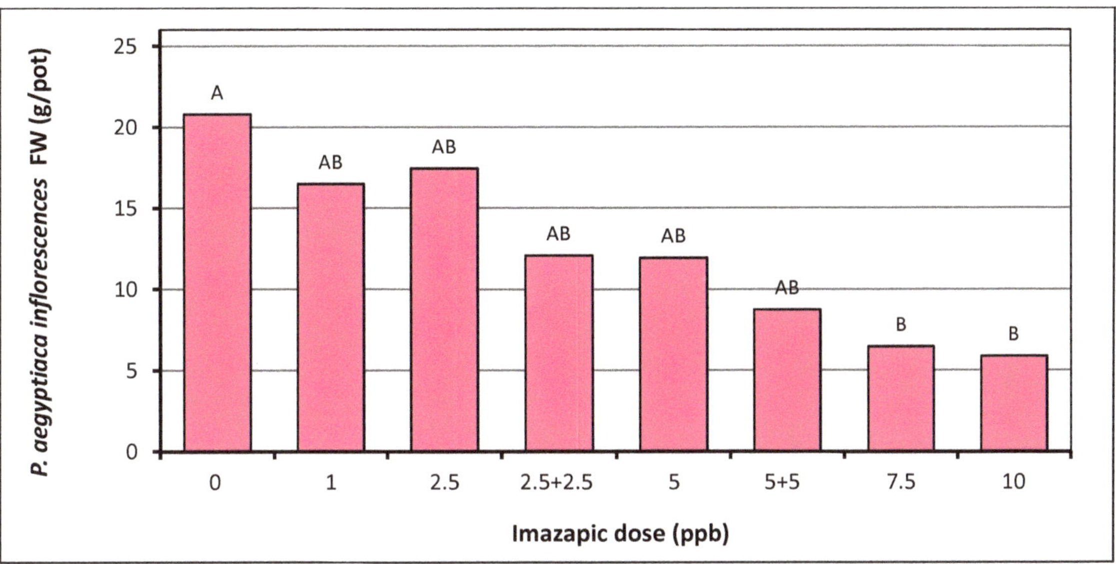

Figure 5. *P. aegyptiaca* aboveground fresh weight (FW) at the end of the imazapic dose-response experiment in pots in the greenhouse, as recorded 71 days after tomato planting. See Table 2 for the description of the different treatments. Letters above bars represent the statistical differences between the *P. aegyptiaca* inflorescences of each treatment analyzed by applying the Student's t LSMeans Differences, $\alpha = 0.05$.

Table 5. Imazapic accumulation in the soil and tomato roots in the pot experiment. The data is averaged over four replications per treatment and $\pm$ represents the standard error of the means of each treatment.

Days after Treatment	Imazapic Soil Concentration (ppb w/w)	Imazapic Tomato Root Concentration (ppb w/w)
1	7.5 ± 2.4	263.6 ± 30.5
3	2.6 ± 0.1	344.3 ± 93.9
7	0.4 ± 0.2	234.6 ± 87.3

4. Discussion and Conclusions

The laboratory and greenhouse experiments demonstrate the complexity of chemical control of broomrape via soil drench/chemigation with imazapic. Successful control depends on many factors, and the interactions among them: herbicide, soil, host plant, parasitic-plant, temperature and moisture. In the presented studies, some of these factors were examined under controlled conditions, enabling the transfer of efficient and selective *P. aegyptiaca* control strategies under field conditions.

In the Petri dish experiments, we elucidated that broomrape seeds and seedlings that are not attached to host plant roots are not sensitive to imazapic even at extreme concentrations. The observation in the PE bags supported this finding by showing that *P. aegyptiaca* is injured by imazapic only after attachment to the tomato host root at imazapic concentrations of 2.5 and 5 ppb in the culture solution.

Imazapic is an ionizable herbicide; thus, it can exist in the soil in two different forms depending on the pH. When the pH is below 3.6, the molecular form predominates, and as the pH increases the anion prevails and the attractive forces between the herbicide and the soil components decrease. The result is that imazapic exhibits extremely low sorption behavior in most agricultural soils [16]. The Koc values for imazapic in the four soils studied are fairly uniform. The fact that imazapic is present mainly in the ionized form in these soils due to their high pH values should rule out hydrophobic sorption by the soil organic matter (SOM) as the main mechanism of sorption. Sorption of the anionic form of imazapic may occur on the broken edges of clay minerals or on sesqui-oxides in the soils. In any case, the extremely low sorption values (Kd) are what determine the transport of imazapic in soils and they indicate that imazapic applied via drip irrigation will be highly mobile in the soil. Thus, when applied at a constant concentration the entire wetted volume of the soil will contain imazapic with the exception of a small volume at the edges of the wetted front. Subsequent irrigation without imazapic will leach the chemical to the boundaries of the wetted zone leaving an imazapic-free volume around the emitter. Similarly, if imazapic is applied as a pulse at the beginning of the irrigation cycle, it will be leached out the volume around the emitter. Similar behavior was observed for bromacil, a slightly adsorbed herbicide, in various soils [19,20]. The differences in the Kd values of imazapic in different soils as elucidated in this study can explain differences in the efficacy of the herbicide in control of *P. aegyptiaca* and has to be taken into account when devising a soil-applied herbicide-based control scheme.

In the greenhouse experiments, we elucidated that imazapic can selectively control *P. aegyptiaca* attached to tomato plant roots, but effective control requires frequent repeated applications to maintain this concentration in the soil throughout the tomato plant growth period.

The tracking of imazapic in soil and tomato roots showed that imazapic concentration in the soil rapidly diminishes, such that 7 DAA very little herbicide remains in the soil, thus requiring repeated herbicide applications. The good tolerance of the tomato host and the high phytotoxicity of imazapic to the attached parasite are due to a source-sink relationship in which the parasite acts as a strong sink absorbing most of the herbicide and thus relieving the tomato plant from phytotoxic effects. The imazapic dissipation mechanism in the soil is most probably due to microbial metabolism [17].

Numerous control methods have been attempted to effectively and selectively control broomrape under field conditions, but most of these attempts have resulted in limited or partial control. The presented laboratory and greenhouse studies have provided invaluable knowledge for devising field imazapic application strategies via soil drench or drip irrigation systems for efficient and selective broomrape control and supported the successful development of field experiments and Decision Support Systems as described by Eizenberg et al. [14], Ephrath et al. [21] and Eizenberg and Goldwasser [13].

Author Contributions: Conceptualization, Y.G., O.R., Z.G. and B.R.; methodology, Y.G., O.R., Z.G., M.S., A.N., A.P., H.K. and B.R.; software, Y.G. and O.R.; validation, Y.G., O.R., Z.G. and B.R.; formal analysis- A.N. investigation, All; resources, A.N.; data curation, A.N.; writing—original draft preparation, Y.G, O.R., Z.G. and B.R.; writing—review and editing, Y.G, O.R., Z.G. and B.R.; visualization, A.N.; supervision, A.N.; project administration, A.N.; funding acquisition, B.R. and Z.G. All authors have read and agreed to the published version of the manuscript.

Funding: This study was part of the National Project for Alleviating Broomrape Damages in Vegetables funded by the Chief Scientist of the Israeli Ministry of Agriculture and Rural Development (Project No. 132-1499-11).

Institutional Review Board Statement: Not applicable.

Informed Consent Statement: Not applicable.

Data Availability Statement: The data is contained within the article.

Acknowledgments: We are grateful for the valuable help of the following people and institutions: The Rubin lab—HUJI, Israel; Hanan Eizenberg and the Weed Research Department at Newe Yaar Research Station, Ramat Yishay, Israel; The Zev Gerstl lab at the Institute of Soil, Water and Environmental Sciences, Volcani Center, ARO, Bet Dagan, Israel; Hillel Manor and 'Hishtil' Israel.

Conflicts of Interest: The authors declare no conflict of interest.

References

1. Heide-Jorgensen, H.S. Introduction: The parasitic syndrome in higher plants. In *Parasitic Orobanchaceae*; Joel, D.M., Gressel, J., Musselman, L.J., Eds.; Springer: Berlin/Heidelberg, Germany, 2013; pp. 1–18.
2. Goldwasser, Y.; Rodenberg, J. Integrated Agronomic Management of Parasitic Weed Seed Banks. In *Parasitic Orobanchaceae*; Joel, D.M., Gressel, J., Musselman, L.J., Eds.; Springer: Berlin/Heidelberg, Germany, 2013; pp. 393–413.
3. Parker, C. The Parasitic Weeds of the Orobanchaceae. In *Parasitic Orobanchaceae*; Joel, D., Gressel, J., Musselman, L., Eds.; Springer: Berlin/Heidelberg, Germany, 2013. [CrossRef]
4. Parker, C.; Riches, C. *Parasitic Weeds of the World: Biology and Control*; CAB International: Wallingford, UK, 1993.
5. Goldwasser, Y.; Kleifeld, Y. Recent approaches to *Orobanche* management—A review. In *Weed Biology and Management*; Inderjit, Ed.; Kluwer Academic Publishers: Dordrecht, The Netherlands, 2004; pp. 439–466.
6. Joel, D.M.; Gressel, J.; Musselman, L.J. *Parasitic Orobanchaceae*; Springer: Berlin/Heidelberg, Germany, 2013; 513p. [CrossRef]
7. CABI Invasive Species Compendium-Orobanche aegyptiaca. Available online: https://www.cabi.org/isc/datasheet/37742#toidentity (accessed on 23 May 2021).
8. Miyao, G. Egyptian broomrape eradication effort in California: A progress report on the joint effort of regulators, university, tomato growers and processors. In Proceedings of the XIV International Symposium on Processing Tomato, ISHS Acta Horticulturae 1159, Santiago, Chile, 6–9 March 2016. [CrossRef]
9. Aparicio, M.F.; Delavault, P.; Timko, M.P. Management of infection by parasitic weeds: A review. *Plants* **2021**, *9*, 1184. [CrossRef] [PubMed]
10. Bai, J.; Wei, Q.; Shu, J.; Gan, Z.; Li, B.; Yan, D.; Huang, Z.; Guo, Y.; Wang, X.; Zhang, L.; et al. Exploration of resistance to *Phelipanche aegyptiaca* in tomato. *Pest Manag. Sci.* **2020**, *76*, 3806–3821. [CrossRef] [PubMed]
11. Goldwasser, Y.; Kleifeld, Y.; Plakhine, D.; Rubin, B. Variation in vetch (*Vicia* spp.) response to *Orobanche aegyptiaca. Weed Sci.* **1997**, *45*, 756–762. [CrossRef]
12. Shaner, D.L. *Herbicide Handbook*, 10th ed.; Weed Science Society of America: Champaign, IL, USA, 2014.
13. Eizenberg, H.; Goldwasser, Y. Control of Egyptian Broomrape in Processing Tomato: A Summary of 20 Years of Research and Successful Implementation. *Plant Dis.* **2018**, *102*, 1477–1488. [CrossRef] [PubMed]
14. Eizenberg, H.; Aly, R.; Cohen, Y. Technologies for smart chemical control of broomrape (*Orobanche* spp. and *Phelipanche* spp.). *Weed Sci.* **2012**, *60*, 316–323. [CrossRef]
15. Hershenhorn, Y.; Eizenberg, H.; Dor, E.; Kapulnik, Y.; Goldwasser, Y. *Phelipanche aegyptiaca* management in tomato. *Weed Res.* **2009**, *49* (Suppl. 1), 34–37. [CrossRef]
16. Lewis, K.A.; Tzilivakis, J.; Warner, D.; Green, A. An international database for pesticide risk assessments and management. *Hum. Ecol. Risk Assess. Int. J.* **2016**, *22*, 1050–1064. [CrossRef]
17. Tu, M.; Hurd, C.; Randall, J.M. *Weed Control Methods Handbook*; The Nature Conservancy: Arlington County, VA, USA, 2001; Available online: http://tncweeds.ucdavis.edu (accessed on 8 June 2021).
18. American Cyanamid Company. *Plateau Herbicide, for Weed Control, Native Grass Establishment and Turf Growth Suppression on Roadsides and Other Non-Crop Areas, PE-47015*; American Cyanamid Company: Parsippany, NJ, USA, 2000.
19. Gerstl, Z.; Yaron, B. Behavior of bromacil and napropamide in soils. II. Distribution after application from a point source. *Soil Sci. Soc. Am. J.* **1983**, *47*, 478–483. [CrossRef]
20. Gerstl, Z.; Albasel, N. Field distribution of pesticides applied via a drip irrigation system. *Irrig. Sci.* **1984**, *5*, 181–193. [CrossRef]
21. Ephrath, J.E.; Hershenhorn, J.; Achdari, G.; Bringer, S.; Eizenberg, H. Use of logistic equation for detection of the initial parasitism phase of Egyptian broomrape (*Phelipanche aegyptiaca*) in tomato. *Weed Sci.* **2012**, *60*, 57–63. [CrossRef]

Article

Biological and Transcriptomic Characterization of Pre-Haustorial Resistance to Sunflower Broomrape (*Orobanche cumana* W.) in Sunflowers (*Helianthus annuus*)

Dana Sisou [1,2,3,*], Yaakov Tadmor [2], Dina Plakhine [1], Hammam Ziadna [1], Sariel Hübner [4] and Hanan Eizenberg [1]

1. Department of Phytopathology and Weed Research, Agricultural Research Organization, Newe Ya'ar Research Center, Ramat Yishay 30095, Israel; dinap@volcani.agri.gov.il (D.P.); hammam@volcani.agri.gov.il (H.Z.); eizenber@volcani.agri.gov.il (H.E.)
2. Department of Vegetable and Field Crops, Agricultural Research Organization, Newe Ya'ar Research Center, Ramat Yishay 30095, Israel; tadmory@volcani.agri.gov.il
3. The Robert H. Smith Institute of Plant Sciences and Genetics, The Robert H. Smith Faculty of Agriculture, Food and Environment, The Hebrew University of Jerusalem, Rehovot 7610001, Israel
4. Galilee Research Institute (MIGAL), Tel-Hai Academic College, Upper Galilee 11016, Israel; sarielh@migal.org.il
* Correspondence: dana.sisou@mail.huji.ac.il; Tel.: +972-50-7712451

Citation: Sisou, D.; Tadmor, Y.; Plakhine, D.; Ziadna, H.; Hübner, S.; Eizenberg, H. Biological and Transcriptomic Characterization of Pre-Haustorial Resistance to Sunflower Broomrape (*Orobanche cumana* W.) in Sunflowers (*Helianthus annuus*). Plants **2021**, 10, 1810. https://doi.org/10.3390/plants10091810

Academic Editors: Bartosz Jan Płachno and Pablo Castillo

Received: 19 July 2021
Accepted: 17 August 2021
Published: 30 August 2021

Publisher's Note: MDPI stays neutral with regard to jurisdictional claims in published maps and institutional affiliations.

Abstract: Infestations with sunflower broomrape (*Orobanche cumana* Wallr.), an obligatory root parasite, constitute a major limitation to sunflower production in many regions around the world. Breeding for resistance is the most effective approach to reduce sunflower broomrape infestation, yet resistance mechanisms are often broken by new races of the pathogen. Elucidating the mechanisms controlling resistance to broomrape at the molecular level is, thus, a desirable way to obtain long-lasting resistance. In this study, we investigated broomrape resistance in a confectionery sunflower cultivar with a robust and long-lasting resistance to sunflower broomrape. Visual screening and histological examination of sunflower roots revealed that penetration of the broomrape haustorium into the sunflower roots was blocked at the cortex, indicating a pre-haustorial mechanism of resistance. A comparative RNA sequencing between broomrape-resistant and -susceptible accessions allowed the identification of genes that were significantly differentially expressed upon broomrape infestation. Among these genes were β-1,3-endoglucanase, β-glucanase, and ethylene-responsive transcription factor 4 (ERF4). These genes were previously reported to be pathogenesis-related in other plant species. This transcriptomic investigation, together with the histological examinations, led us to conclude that the resistance mechanism involves the identification of the broomrape and the consequent formation of a physical barrier that prevents the establishment of the broomrape into the sunflower roots.

Keywords: sunflower (*Helianthus annuus*); broomrape (*Orobanche cumana*); broomrape resistance; transcriptomics; parasitic plants

1. Introduction

Among the plethora of plant pathogens, parasitic weeds are considered a major threat to crops worldwide. Broomrape species (*Orobanche* and *Phelipanche* spp., Orobanchaceae) are obligatory parasitic plants that are particularly damaging to agricultural crops, especially legumes, tobacco, carrot, tomato, and sunflower (*Helianthus annuus* L.). Sunflower broomrape (*Orobanche cumana* Wallr.) thus constitutes a major constraint on sunflower production in many regions around the globe, including the Middle East, Southeast Europe, Southwest Asia, Spain, and China [1]. Because broomrape is a chlorophyll-lacking holoparasite, it obtains all its nutritional requirements from the host plant. The parasitism occurs at the host roots, damaging host development and resulting in significant yield reduction [2]. Broomrape control is a challenging problem because only a few herbicides are effective against broomrape and, more importantly, because the parasite's attachment to the host root tissues allows systemic herbicides to move from the parasite into the host [2,3]. Therefore, breeding for resistant varieties is the most efficient and sustainable means to control broomrape in sunflower. Generally, there are three types of host resistance to broomrape, in accordance with the developmental stage of the parasitism: The first, a pre-attachment resistance mechanism, depends on the ability of the host to prevent the attachment of the parasite, including the prevention of parasite germination and development, as well as low production or release of germination stimulants [4] such as strigolactones from the host roots into the rhizosphere [5,6]. If pre-attachment resistance fails, broomrape seeds will germinate, and the parasites will grow toward the host roots via chemotropism and attach to the roots [7]. The second resistance mechanism—known as post-attachment or pre-haustorial resistance [4]—is a mechanism inhibiting penetration into the host root cells as well as the development of the haustorium, thus preventing vascular conductivity between the parasite and the host [8]. This resistance involves the production of physical barriers (such as thickening of host root cell walls by lignification and callose deposition) [4,9,10], which prevents the parasite from establishing a vascular connection with the host roots. The third, post-haustorial type of resistance involves the release of a gum-like substance [11,12] and the production and delivery of toxic compounds (phenolics) by the host. The transfer of these chemical compounds to the parasite prevents or delays the formation of the tubercles that are necessary for stalk elongation and flowering of the parasite [11,13,14]. To shed light on the basis of the resistance mechanisms in sunflowers, it is first necessary to understand the structure of the plant innate immunity system. The first level of the plant immune system is pathogen-triggered immunity (PTI), which is activated by the recognition of pathogen-associated molecular patterns (PAMPs). While, over time, pathogens have developed effectors to inhibit the PAMP-activated PTI response, plants, in turn, have evolved to perceive and counteract these effectors through a second layer of defense, known as effector-triggered immunity (ETI), formerly known as gene-for-gene resistance [15]. The rapid changes in the race composition of sunflower broomrape have led to an ongoing gene-for-gene 'arms race' between breeders and the parasitic weed. The development of *O. cumana*-resistant cultivars usually includes the introgression of resistance genes, which are, in many cases, broken by the parasite. This resistance breakdown occurs due to the massive use of vertical (monogenic) resistance [16] and can be addressed by the introduction of horizontal (quantitative) resistance genes with the aim of developing a more durable resistance [17,18]. Several *Orobanche* resistance (Or) QTLs that confer resistance to *O. cumana* have been used in breeding programs over the years [19]. These QTLs were numbered (Or1–Or6) in accordance with the gene-for-gene model [20], and they provide resistance against the corresponding *O. cumana* races A to F [19,21–24]. QTL Or5 was the first to be mapped and was located on the telomeric region of chromosome 3 [25–27]. Following this successful attempt to map Or5, a number of studies used the genetic mapping approach to locate more resistance QTLs in sunflower. For example, Perez-Vich et al. (2004) [17] detected eight QTLs for resistance along seven different chromosomes. More recently, Louarn et al. (2016) [28] studied resistance to races F and G and identified a total of 17 QTLs in accordance with different stages of broomrape development. These results were further supported by Imerovski et al. (2019) [29]. In 2019, Duriez et al. were able to target the first broomrape resistance gene, HaOr7, found on chromosome 7, which encodes a leucine-rich repeat receptor-like kinase [30]. In this study, we examined the resistance of the confectionery hybrid cultivar 'EMEK3' (developed by Sha'ar Ha'amakim Seeds, Ltd.), which has high, long-term resistance to sunflower broomrape, with the aim to elucidate—biologically and transcriptomically—the broomrape resistance mechanism, an essential step toward the development of effective sunflower breeding programs.

2. Results

2.1. Effect of Grafting on the Source of the Resistance

To test whether biological compounds that are produced in above-ground tissues are involved in the resistance to broomrape, a grafting experiment was conducted. All possible combinations of resistant and susceptible rootstalk/scion grafts were generated, non-grafted plants were used as a control, and the development of the broomrape was monitored. Overall, no parasitism was observed on resistant roots or rootstalks regardless of the type of scion used. All plants with a susceptible root or rootstalk were infested with 420–450 *O. cumana* tubercles and stalks of different sizes, regardless of whether the grafted scions were resistant or susceptible (Figure 1a,b). These results indicate that the aboveground tissues of the sunflower plant do not contribute substantially to its resistance to broomrape.

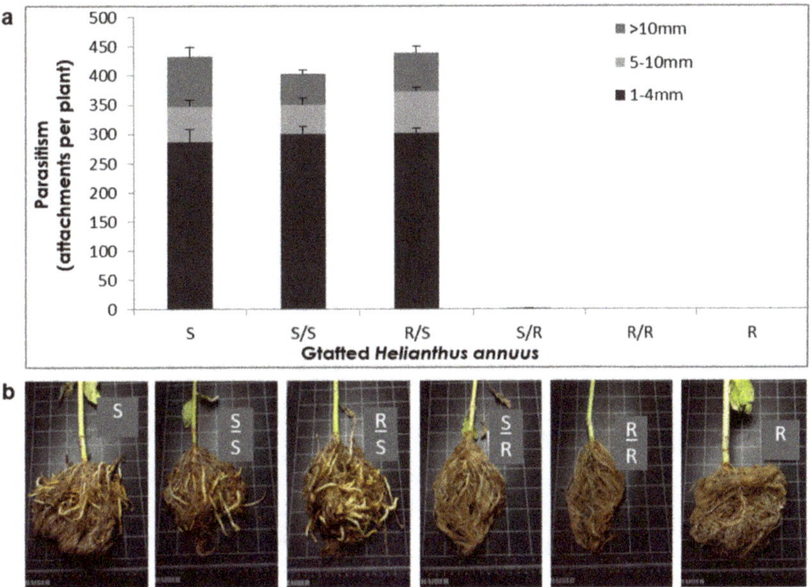

Figure 1. (**a**) Number of *O. cumana* tubercles parasitizing grafted sunflower plants (+SE). S: non-grafted susceptible sunflower; S/S: self-grafted susceptible sunflower; R/S: resistant sunflower shoot grafted onto susceptible sunflower rootstock; S/R: susceptible sunflower shoot grafted onto resistant sunflower rootstock; R/R: self-grafted resistant sunflower; R: non-grafted resistant sunflower. (**b**) Grafted and non-grafted sunflower roots 52 d post-infestation with *O. cumana*.

2.2. Sunflower–O. cumana Incompatibility

Two key parasitism stages were monitored periodically: germination and attachments (Figure 2a,b). The germination rate of *O. cumana* seeds was higher (50%) in the presence of the resistant cultivar roots than in the presence of the susceptible cultivar roots (39%) (Figure 2b). The first *O. cumana* attachment was observed 11 d after infestation in the susceptible cultivar, while no attachments were observed in the resistant cultivar (Figure 2a). The observation of *O. cumana* seedlings growing together with sunflower plantlets in clear polyethylene bags (PEB) revealed that the development of the *O. cumana* seedlings was arrested after attaching and attempting to invade the roots of the resistant cultivar, thereby preventing parasite establishment. The disruption of the parasite penetration into the host roots and the subsequent deterioration of the parasite seedlings was accompanied by a darkening of host and parasite tissues at the penetration point (Figure 3a,b). In contrast, establishment and development of healthy tubercles was observed in the roots of the susceptible cultivar (Figure 3c,d). The time at which the resistance response was induced most strongly after the attachment of the parasite seedling to the host roots was determined by observing the host–parasite system growing in the PEB system for 21 d. A large increase in necrotic *O. cumana* seedlings in the presence of the resistant cultivar roots was observed five d after infestation; at this time point, the necrosis of *O. cumana* seedlings that had attached to the resistant cultivar roots grew from 0 to 44% (percentage of germinated seeds), while on the susceptible cultivar roots, only 9% appeared necrotic (Figure S1). Histological examination of 'EMEK3' (resistant) and 'D.Y.3' (susceptible) roots along with the attached parts of *O. cumana* seedlings, which were sampled five d post infection, showed that the intruding *O. cumana* cells were blocked at the cortex of the resistant cultivar roots and could not reach the endodermis (Figure 4a,b). The root endodermal cells of the resistant cultivar and the attached *O. cumana* seedling-intruding cells were stained with safranin, indicating lignification of cell walls, which presumably prevented the connection of the parasite to the host vascular system and hence, the development of the parasite, whereas in the susceptible cultivar, the formation of the haustorium and compatible connection to the vascular system was observed (Figure 4c,d).

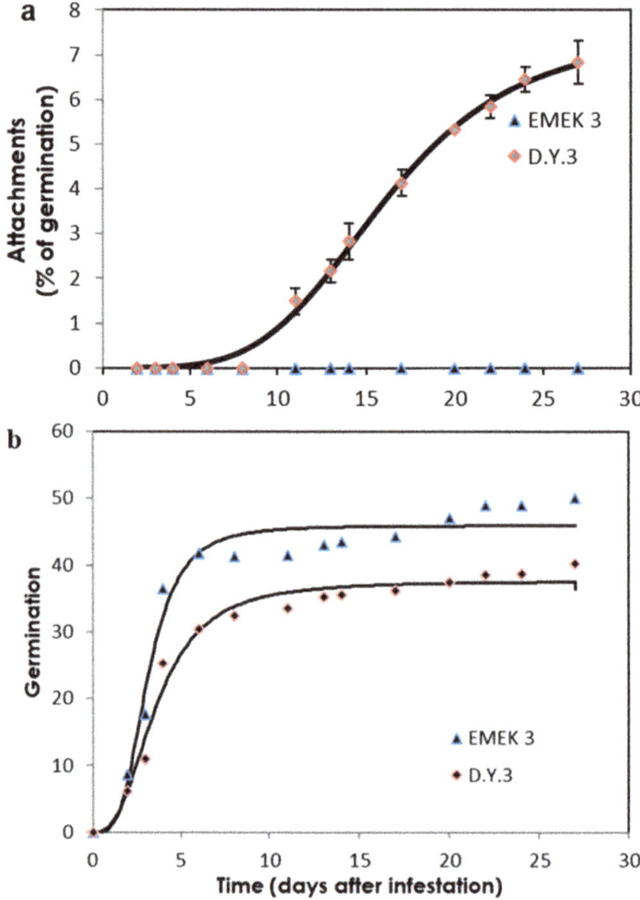

Figure 2. Parasitism dynamics of *O. cumana* on resistant ('EMEK3') and susceptible ('D.Y.3') sunflowers grown in a polyethylene bag system. Attachment (% of germinated seeds) (**a**) and germination (**b**) of *O. cumana* in the presence of resistant and susceptible sunflower cultivars.

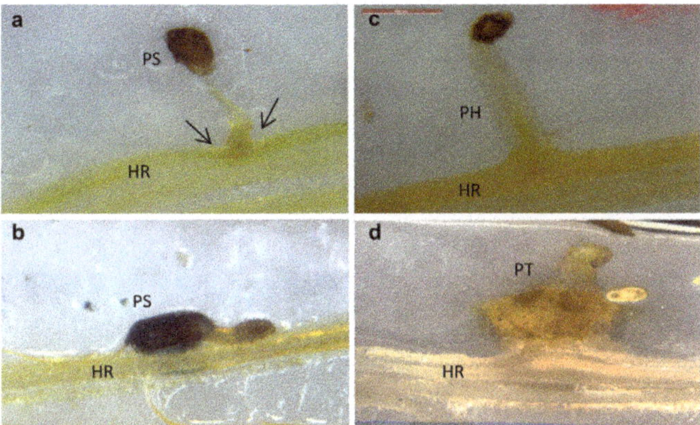

Figure 3. Resistant ('EMEK3') (**a,b**) and susceptible ('D.Y.3') (**c,d**) sunflower roots infested with *O. cumana*, 10 (**a,c**) and 21 (**b,d**) d post infestation. PH: parasite haustorium; HR: host root; PS: parasite seedling; PT: parasite tubercle.

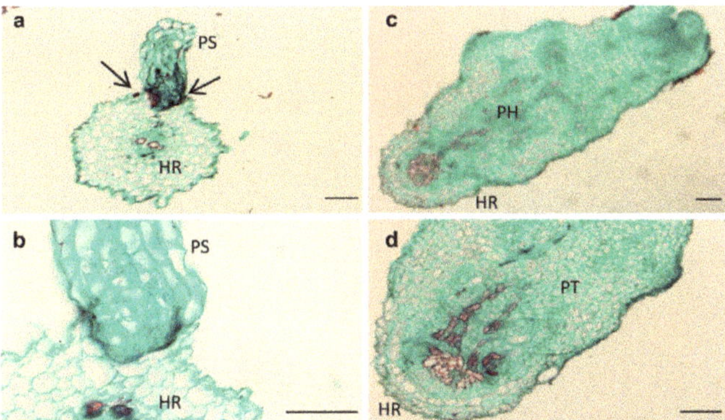

Figure 4. Cross-sections of compatible and incompatible interactions of *O. cumana* with resistant ('EMEK3') (**a,b**) and susceptible ('D.Y.3') (**c,d**) sunflower roots five d post-infestation. PH: parasite haustorium; HR: host root; PS: parasite seedling; PT: parasite tubercle. Scale bar = 100 μm.

2.3. Identification of Candidate Resistance Genes, Using RNA-Sequencing

Comparative RNA-Seq of *O. cumana*-infested and -non-infested sunflower roots of the resistant (E) cultivar, the R bulk, and the S bulk were used to identify differentially expressed genes (DEGs) associated with sunflower resistance. A total of 7.4–9.8 $\times$ 10^6 reads were produced with an average of 8,648,866.3 reads per library.

Out of 1123 and 348 genes that were differentially expressed pre-infestation and five d post-infestation in the R bulk and 'EMEK3', respectively, 37 genes were found to be communal and not differentially expressed in the S bulk (Figure 5a). To exclude genes that were not related to broomrape infestation, we cross-compared the DEGs of non-infested samples collected on the infestation day and at five d post-infestation: 47 genes were found to be communal to R bulk and 'EMEK3' (Figure 5b). These 47 genes were then cross-compared with the 37 genes previously mentioned. Two genes were found to be communal and were therefore discarded (Figure 5c). Hence, 35 genes were classified as related to the resistance response (Figure 5c). Thereafter, we cross-compared the genes that were differentially expressed in 'EMEK3' among all treatments: because there was a five-day difference between sampling dates, we assumed that some of the DEGs were not related to the resistance response (i.e., regulatory genes). Therefore, we focused on the

44 genes that were differentially expressed pre-infestation and at the time corresponding to five d post-infestation with and without *O. cumana* (Figure 5d). Finally, these 44 DEGs were cross-compared with the 35 DEGs communal to R bulk and 'EMEK3' during the resistance response; 3 genes were found to be mutual (Figure 5e). These three genes were annotated to the sunflower genome and were identified as β-glucanase, β-1,3-endoglucanase, and ethylene-responsive transcription factor 4 (ERF4). The expression levels of the genes encoding β-1,3-endoglucanase and β-glucanase were 2.49 and 2.5 times higher in 'EMEK3' roots, respectively (Figure 6a,b). The expression level of the gene encoding ERF4 was 2.97 times lower in 'EMEK3' roots five d after the infestation with *O. cumana* (Figure 6c). Gene ontology (GO) enrichment analysis, performed on 1439 significant DEGs in the resistant cultivar, showed that 224 overexpressed genes were significantly enriched (false discovery rate (FDR) < 0.05). The most enriched term in the *Biological Process* class was "metabolic process" (74%). For the *Molecular Function* and *Cellular Component* classes, these terms were "catalytic activity" (67%) and "cell periphery" (14%), respectively (Figure 7).

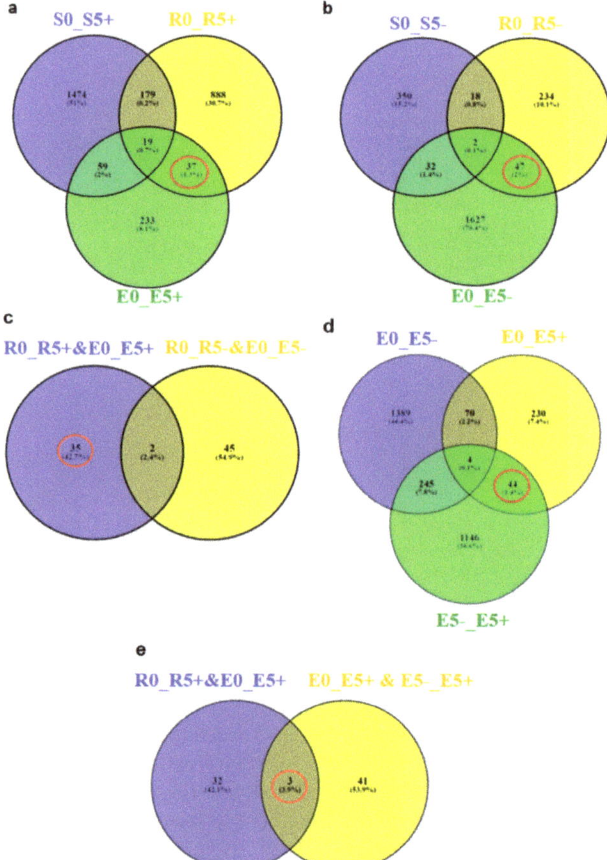

Figure 5. Venn diagram of the DEGs between the R bulk (R), S bulk (S), and EMEK3 (E) pre-infestation (0), five d post-infestation with *O. cumana* (5+) (**a**) and five d post-infestation without *O. cumana* (5−) (**b**). (**c**) Venn diagram of the communal DEGs of (**a,b**). (**d**) The DEG in EMEK3 pre- and five d post-infestation with or without *O. cumana*. (**e**) Venn diagram of the communal DEGs of (**c,d**).

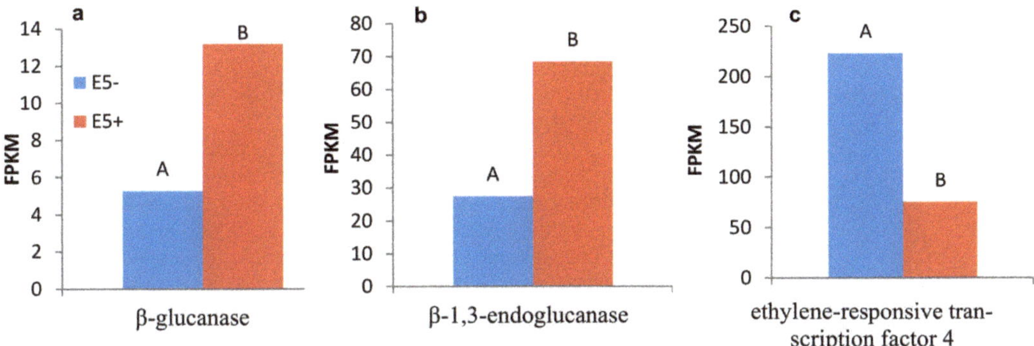

Figure 6. Expression levels (fragments per kilobase million) of the genes encoding β-glucanase (**a**), β-1,3-endoglucanase (**b**), and ethylene-responsive transcription factor 4 (**c**) in EMEK3 roots five d post-infestation with *O. cumana* (E5+) and in non-infested (E5−) roots. Different letters indicate significant difference between groups.

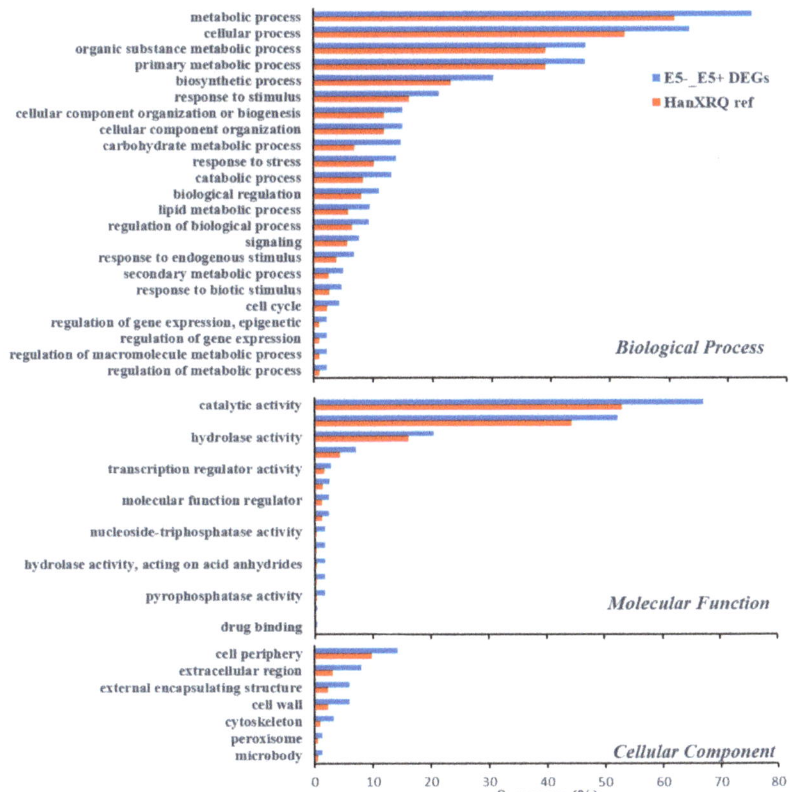

Figure 7. Distribution of enriched GO terms for differentially over-expressed genes (Fisher's Exact Test) for DEGs of the resistant cultivar EMEK3 [i.e., genes that were differentially expressed between roots of EMEK3 five d post infestation with *O. cumana* (E5+) and non-infested roots (E5-)] compared with GO terms of whole reference-predicted gene annotation (HanXRQ). The Y-axis represents significant enrichment of GO terms and the X-axis shows the relative frequency of the term.

3. Discussion

A variety of strategies comprising the host defense response at the early stages of the parasite life cycle have been described in a number of studies, namely, the lignification and subarization of host cell walls [10]; the accumulation of callose, peroxidases, and H_2O_2 in the cortex and protein cross-linking in the cell walls [11,12]; phenylalanine ammonia lyase (PAL) activity and high concentrations of phenolic compounds in the host roots [31–33]; and degeneration of tubercles after establishment [14]. Determining the phenological stage at which the incompatibility occurs is crucial for understanding the resistance mechanism and the molecular basis that governs it. To this end, we set up an observation system based on transparent PEBs that enabled us to follow the sunflower–broomrape interaction continuously and thereby to overcome the difficulty of detecting the exact time at which response was maximal. Our observations revealed that the resistance response began in the early stages of the parasite life cycle, after germination and attachment to the roots, and while the parasite was attempting to penetrate into the host roots (Figures 2–4). Blocking of the penetration attempt was accompanied by the necrosis of parasite and host tissues in the penetration area, suggesting a pre-haustorial mechanism of resistance [12]. Our PEB system showed a markedly high necrosis rate of the attached broomrape seedlings in the roots of the resistant cultivar at five d post-infestation (Figure S1). This high death rate was attributed to the prevention of penetration into the host roots and hence, prevention of the establishment in the vascular system that is vital for the parasite seedlings. We thus confirmed by histological methodologies that the parasite intrusion was blocked in the host cortex before the parasite could reach the host endodermis. The endodermal cells in the vicinity of the intrusive broomrape cells in the penetration area were colored with safranin, indicating the involvement of lignin in the host response (Figure 4). Suberization, lignification, and cell wall thickening have previously been ascribed to the sunflower defense response to *O. cumana* [10,34,35]. We excluded the possibility that the host shoot was involved in the resistance response by grafting susceptible sunflower scions onto resistant rootstocks and vice versa. The resistant cultivar rootstocks conferred resistance on the susceptible scions, but the susceptible rootstocks were parasitized with *O. cumana* regardless of whether the grafted scions were resistant or susceptible (Figure 1). Similar results have been obtained for the resistance of the tomato (*Solanum lycopersicum*) to several broomrape species, namely, *P. aegyptiaca*, *P. ramosa*, *O. cernua*, and *O. crenata* [36], implying that the resistance response is expressed exclusively in the roots. A comparative transcriptome analysis of infested and non-infested resistant and susceptible sunflower roots detected 1439 significant DEGs in the roots of the resistant cultivar post-infestation. GO enrichment analyses of these DEGs were performed to infer the biological processes and the functions of the genes associated with the resistance response, with the ontology analysis revealing a number of overexpressed GO terms (Figure 7). Importantly, terms associated with the cell periphery (14%), the extracellular region (7.9%), the external encapsulating structure (5.9%), and the cell wall (5.9%) were significantly enriched in the *Cellular Component* category, indicating high activity in these regions. The *Biological Process* category included response to stimulus (21%), cellular component organization (15%), and response to stress (13.8%) (Figure 7). Finally, a series of Venn diagrams [37] facilitating cross-comparisons of DEGs in the R bulk, the S bulk, and the resistant cultivar 'EMEK3' before and after infestation with *O. cumana* identified three genes that were differentially expressed between infested and non-infested sunflower roots of both 'EMEK3' and the R bulk (Figure 5). As a consequence of the infestation, two of these genes, β-1,3-endoglucanase and β-glucanase, were upregulated, and the third gene, ERF4, was downregulated. These findings indicate activation of the plant's innate immune system, in which the recognition of PAMPs activates a hypersensitive response and the accumulation of pathogenesis-related (PR) proteins [38,39] such as β-glucanases, which are PR proteins belonging to the PR-2 family. This family of proteins is believed to play an important role in plant defense responses to pathogen infection [40–42]. Indeed, it has been shown that β-glucanases, which are able to degrade cell wall β-glucan, are involved in resistance to

O. crenata in peas (*Pisum sativum*) [11,43] and in sunflower resistance to *O. cumana* [33]. Downregulation of the ERF4 gene post-infection should be viewed in the context of the role of the endogenous hormone, ethylene, in regulating defense responses in plants, including the regulation of gene expression during adaptive responses to abiotic and biotic stresses [44]. The ERF transcription factors, which are unique to plants, have a binding domain that can bind to the GCC box, an element found in the promoters of many defense, stress-responsive, and PR genes [45,46]. Just as there is a range of stresses, there are a large a number of ERFs, with many of the ERFs being transcription activators. Indeed, AtERF1, AtERF2, and AtERF5 act as transcriptional activators, although AtERF3 and AtERF4 act as transcriptional repressors for GCC box-dependent transcription in Arabidopsis leaves [47]. In that context, McGrath et al. (2005) [48] demonstrated that the Arabidopsis *erf4-1* mutant was resistant to *Fusarium oxysporum*, while transgenic lines overexpressing AtERF4 were susceptible, and therefore concluded that AtERF4 negatively regulates resistance to *F. oxysporum*. The downregulation of the ERF4 gene in the roots of the resistant sunflower post-*O. cumana* infestation suggests that, as in Arabidopsis, the sunflower response to biotic stress is negatively regulated by ERF4. Furthermore, the recent study of Liu et al. (2020) [49], using bulked segregant RNA-Seq (BSR-Seq), identified ERF as a candidate gene for *O. cumana* resistance in sunflowers. Taken together, the results obtained for the biological characterization combined with those for the genetic characterization provide a comprehensive view of the relations between the resistant cultivar 'EMEK3' and *O. cumana*. This broad view allowed us to propose the following resistance mechanism model: After 'EMEK3' induces *O. cumana* seed germination, the seedlings' attachment to the sunflower roots is perceived by PAMPs. These molecules set off a PTI response that downregulates ERF and abrogates the suppression of PR genes (including β-glucanase). β-glucanase then breaks down the parasite cell walls which, in turn, release effectors that trigger the second level of the plant immune response, namely, effector-triggered immunity (ETI). As a result, a physical barrier is created by the accumulation of lignin and other phenolic compounds in the penetration area, and the *O. cumana* seedlings fail to establish a connection with the host vascular system, leading to parasite necrosis.

4. Conclusions

Sunflower pre-haustorial resistance to sunflower broomrape involves the expression of β-1,3-endoglucanase, β-glucanase, and ethylene-responsive transcription factor 4 (ERF4) genes. The resistance mechanism includes the identification of the broomrape and the formation of a physical barrier that prevents the penetration of the broomrape into the sunflower roots.

5. Materials and Methods

5.1. Plant Materials and Growth Conditions

The cultivars 'EMEK3' (resistant) and 'D.Y.3' (susceptible) as well as 12 sunflower breeding accessions that are being used in breeding programs for the introgression of different traits for the development of Israeli sunflower cultivars were kindly provided by Sha'ar Ha'amakim Seeds, Ltd. (Sha'ar Ha'amakim, Israel). *O. cumana* inflorescences were collected from an infested sunflower field in northern Israel in 2012. The seeds were separated from the capsules, using 300-mesh sieves, and stored in the dark at 4 °C prior to use. The germination rate of these *O. cumana* seeds at 25 °C was 85%.

5.2. Preconditioning of O. cumana Seeds

Preconditioning was performed under sterile conditions. The seeds were surface sterilized for 2.5 min in ethanol (70%) followed by 10 min in sodium hypochlorite (1%) and then rinsed 5 times with sterile distilled water and dried for 2 h in a laminar airflow cabinet. The dried seeds were spread on 5.5 cm diameter glass fiber filter paper discs (Whatman #3, Whatman International, Ltd., Maidstone, England) that had been wetted with 600 μL of sterile distilled water. The discs were placed in sterile, 5.5 cm diameter petri dishes.

The petri dishes were sealed with Parafilm and incubated at 25 °C for 7 d in the dark. Thereafter, 220 µL (10^{-5} M) of GR24 (a commonly used broomrape synthetic germination stimulant [6]) were added to the discs, and the petri dishes were resealed and kept in the dark for another 24 h.

5.3. Cultivation in Polyethylene Bag

The polyethylene bag (PEB) system of Parker and Dixon (1983) [50], with the slight modification of Eizenberg et al. (2003) [51] to tailor it to sunflower cultivation, was used for observing the sunflower–broomrape interaction, as follows: Sunflower seedlings at the cotyledon stage were placed on 25 × 10 cm glass microfiber filter papers (Whatman GF/A), which were then inserted into clear PEBs (35 × 10 cm) and allowed to grow in a growth chamber under controlled conditions (25 °C; 18 h light; 150–200 µE m^{-1} s^{-1}) for 10 d. Approximately 5 µg of preconditioned *O. cumana* seeds were then carefully placed alongside the sunflower roots on the GF/A filter papers. Sterilized, half-strength Hoagland nutrient solution [52] (5 mL) was supplied every day. Observations were carried out every 2–3 d with an electronic binocular microscope (Leica M80) to monitor seed germination, attachment, and establishment or necrosis of the *O. cumana* seedlings.

5.4. Histological Analysis

Plant material for histological analysis was taken from the PEB system. 'EMEK3' and 'D.Y.3' roots, along with the attached parts of *O. cumana* seedlings, were sampled five d post-infection. In parallel, non-infected sunflower roots (control) were sampled. The sampled roots (with and without *O. cumana*) were fixed in FAA (5% formalin:5% acetic acid:90% alcohol (70%), *v/v*) for 24 h. Fixed samples were then dehydrated in an ethanol series (50, 70, 90, 95, 100%; 1–2 h each). After dehydration, the samples were infiltrated with a series of Histo-Clear:ethanol (1:3, 1:1, 3:1 ratio; 1 h each), cleared with Histo-Clear (xylene substitute), and embedded in paraffin. The samples were then cut into 13µm sections with a rotary microtome (Leica RM2245, Leica Biosystems, Nussloch, Germany) and stained with safranin/fast green [53].

5.5. Grafting Experiments

To assess the involvement of the shoot in the resistance mechanism, grafting experiments were conducted as follows: 'EMEK3' and 'D.Y.3' seeds were sown in 2-liter pots, and 14 d post-emergence, the stems of plants with two true leaves were cut above the cotyledons at a 45° angle. 'EMEK3' shoots were grafted onto 'D.Y.3' rootstock and vice versa. The grafted sunflowers were kept in a closed chamber with 100% humidity at 25 °C for 3 d. The plants were then transferred to a humid chamber (in which water was sprayed every 3 h for 10 s) for 7 d. Thereafter, the plants were covered with polyethylene and were gradually exposed to the room atmosphere by removing portions of the polyethylene cover every 1–2 d along 7 d. Once plants were acclimated, they were planted in 2-liter pots and infested with 15 ppm of *O. cumana* seeds. Self-grafted and non-grafted plants were used as a control in the experiment.

5.6. Statistical Analysis

The experiments were carried out in 5 replications in a fully randomized design. The analysis of variance (ANOVA) was performed, and means were compared using the Student's t test ($p < 0.05$) in JMP PRO 12 software (v5.1; SAS Institute, Inc., Cary, NC, USA).

5.7. Bulk Construction and RNA Extraction

Twelve sunflower breeding accessions that had been used as a genetic source for 'EMEK3' breeding were quantified for *O. cumana* resistance under conditions of artificial infestation in pots held in a greenhouse (25–30 °C). Five accessions showed complete resistance with no attachments on the roots, and seven accessions exhibited susceptibility at all *O. cumana* parasitism stages (Figure S2). Therefore, in addition to the resistant cultivar, five resistant accessions and five susceptible accessions were selected to construct a resistant (R) bulk and a susceptible (S) bulk for RNA sequencing. Roots of PEB-cultured sunflowers of five resistant accessions, five susceptible accessions, and the resistant cultivar were collected on the day of infestation and at five d post-infestation with *O. cumana* for both infected and control plants. Whole roots were ground in liquid nitrogen, and equal amounts of root tissue from each accession of the resistant and the susceptible accessions were taken as R and S bulks. Total RNA was isolated from 27 samples ('EMEK3', R bulk, and S bulk × 3 treatments/sampling time × 3 replicates), using a Spectrum^{TM} Plant Total RNA Kit (Sigma-Aldrich, St. Louis, MO, USA) according to the manufacturer's protocol. RNA quality and integrity were evaluated by Agilent TapeStation 2200 (Agilent Technologies, Santa Clara, CA, USA).

5.8. RNA Sequencing and Mapping

Libraries were prepared using the Genomics in-house protocol for mRNA-seq. Briefly, the polyA fraction (mRNA) was purified from 500 ng of total RNA, followed by fragmentation and the generation of double-stranded cDNA. Next, end repair, a base addition, adapter ligation, and PCR amplification steps were performed. Libraries were evaluated by Qubit (Thermo Fisher Scientific) and TapeStation (Agilent). Sequencing libraries were constructed with barcodes to allow multiplexing of 27 samples in 2 lanes. Approximately 16–20 million single-end 60-bp reads were sequenced per sample on an Illumina HiSeq 2500 V4 instrument. The quality of the raw reads was evaluated using FastQC v.0.11.03 [54], followed by trimming and removal of low-quality reads, using Trimmomatic v.036 [55]. Cleaned reads from each of the 27 libraries were then aligned to the *H. annuus* XRQ v1.0 reference genome [56], using STAR v.2.5.2b [57], and the level of expression of each gene in each library was estimated using RSEM v.1.2.31 [58]. Expression levels were normalized using the number of reads per kilobase per million reads mapped (RPKM) for each transcript.

5.9. RNA-Seq Data Analysis

The RSEM output files were analyzed using R package DESeq2 [59] for differential expression analysis. A pairwise comparisons test was performed between 'EMEK3', R bulk, and S bulk. DEGs were considered as significant at FDR < 0.05 [60]. GO terms were obtained from the heliagene database for XRQ (https://www.heliagene.org/HanXRQ-SUNRISE/, accessed on 27 December 2016), and GO terms enrichment analysis was performed for significant DEGs compared to all other GO terms using the Blast2GO (v5.2.5) analysis tools [61]. Significantly over-represented GO terms were identified using Fisher's exact test at a significance level of FDR < 0.05. GO slim (Blast2GO tool) was performed to reduce the complexity of GO terms for functional analysis of annotated *H. annuus* genes.

Supplementary Materials: The following are available online at https://www.mdpi.com/article/10.3390/plants10091810/s1, Figure S1: Sunflower-*O. cumana* incompatibility, Figure S2: Screening of sunflower breeding ac-cessions for resistance to *O. cumana*.

Author Contributions: Y.T. and H.E. conceived the idea and supervised the project overall. D.S., Y.T., D.P., H.Z. and H.E. designed the experiments. D.S. conducted experiments, analyzed data, and wrote the manuscript. S.H. performed the differential expression analysis. All authors have read and agreed to the published version of the manuscript.

Funding: This research received no external funding.

Data Availability Statement: The data from RNA-seq has been submitted into the NCBI SRA database, and the BioProject accession number is PRJNA706194 (https://www.ncbi.nlm.nih.gov/sra/PRJNA706194, accessed on 27 May 2021).

Conflicts of Interest: The authors declare no conflict of interest.

References

1. Parker, C. The parasitic weeds of the Orobanchaceae. In *Parasitic Orobanchaceae*; Joel, D.M., Gressel, J., Eds.; Springer: Berlin/Heidelberg, Germany, 2013; pp. 313–344.
2. Eizenberg, H.; Hershenhorn, J.; Ephrath, J.H.; Kanampiu, F. Chemical control. In *Parasitic Orobanchaceae*; Joel, D.M., Gressel, J., Eds.; Springer: Berlin/Heidelberg, Germany, 2013; pp. 415–432.
3. Eizenberg, H.; Hershenhorn, J.; Ephrath, J.E. Factors affecting the efficacy of *Orobanche cumana* chemical control in sunflower. *Weed Res.* **2009**, *49*, 308–315. [CrossRef]
4. De Luque, A.P.; Moreno, M.; Rubiales, D. Host plant resistance against broomrapes (*Orobanche* spp.): Defence reactions and mechanisms of resistance. *Ann. Appl. Biol.* **2008**, *152*, 131–141. [CrossRef]
5. Xie, X.; Yoneyama, K.; Yoneyama, K. The Strigolactone Story. *Annu. Rev. Phytopathol.* **2010**, *48*, 93–117. [CrossRef] [PubMed]
6. Yoneyama, K.; Ruyter-Spira, C.; Bouwmeester, H. Induction of germination. In *Parasitic Orobanchaceae: Parasitic Mechanisms and Control Strategies*; Joel, D.M., Gressel, J., Musselman, L.J., Eds.; Springer: Berlin/Heidelberg, Germany, 2013; pp. 167–194. ISBN 978-3-642-38146-1.
7. Joel, D.M.; Bar, H. The seed and the seedling. In *Parasitic Orobanchaceae: Parasitic Mechanisms and Control Strategies*; Joel, D.M., Gressel, J., Musselman, L.J., Eds.; Springer: Berlin/Heidelberg, Germany, 2013; pp. 147–165. ISBN 978-3-642-38146-1.
8. Joel, D.M.; Losner-Goshen, D. The attachment organ of the parasitic angiosperms *Orobanche cumana* and *O. aegyptiaca* and its development. *Can. J. Bot.* **1994**, *72*, 564–574. [CrossRef]
9. Letousey, P.; De Zélicourt, A.; Dos Santos, C.V.; Thoiron, S.; Monteau, F.; Simier, P.; Thalouarn, P.; Delavault, P. Molecular analysis of resistance mechanisms to *Orobanche cumana* in sunflower. *Plant Pathol.* **2007**, *56*, 536–546. [CrossRef]
10. Echevarría-Zomeño, S.; de Luque, A.P.; Jorrín, J.; Maldonado, A.M. Pre-haustorial resistance to broomrape (*Orobanche cumana*) in sunflower (*Helianthus annuus*): Cytochemical studies. *J. Exp. Bot.* **2006**, *57*, 4189–4200. [CrossRef]
11. De Luque, A.P.; González-Verdejo, C.I.; Lozano-Baena, M.-D.; Dita, M.A.; Cubero, J.I.; González-Melendi, P.; Risueño, M.C.; Rubiales, D. Protein cross-linking, peroxidase and β-1,3-endoglucanase involved in resistance of pea against *Orobanche crenata*. *J. Exp. Bot.* **2006**, *57*, 1461–1469. [CrossRef]
12. De Luque, A.P.; Rubiales, D.; Cubero, J.I.; Press, M.C.; Scholes, J.; Yoneyama, K.; Takeuchi, Y.; Plakhine, D.; Joel, D.M. Interaction between *Orobanche crenata* and its Host Legumes: Unsuccessful Haustorial Penetration and Necrosis of the Developing Parasite. *Ann. Bot.* **2005**, *95*, 935–942. [CrossRef]
13. Lozano-Baena, M.-D.; Prats, E.; Moreno, M.T.; Rubiales, D.; de Luque, A.P. *Medicago truncatula* as a Model for Nonhost Resistance in Legume-Parasitic Plant Interactions. *Plant Physiol.* **2007**, *145*, 437–449. [CrossRef] [PubMed]
14. Eizenberg, H.; Hershenhorn, J.; Plakhine, D.; Kleifeld, Y.; Shtienberg, D.; Rubin, B. Effect of temperature on susceptibility of sunflower varieties to sunflower broomrape (*Orobanche cumana*) and Egyptian broomrape (*Orobanche aegyptiaca*). *Weed Sci.* **2003**, *51*, 279–286. [CrossRef]
15. Boller, T.; He, S.Y. Innate immunity in plants: An arms race between pattern recognition receptors in plants and effectors in microbial pathogens. *Science* **2009**, *324*, 742–744. [CrossRef]
16. Molinero-Ruiz, L.; Delavault, P.; Pérez-Vich, B.; Pacureanu-Joita, M.; Bulos, M.; Altieri, E.; Domínguez, J. History of the race structure of *Orobanche cumana* and the breeding of sunflower for resistance to this parasitic weed: A review. *Span. J. Agric. Res.* **2015**, *13*, e10R01-01. [CrossRef]
17. Pérez-Vich, B.; Akhtouch, B.; Knapp, S.J.; León, A.J.; Velasco, L.; Fernández-Martínez, J.M.; Berry, S.T. Quantitative trait loci for broomrape (*Orobanche cumana* Wallr.) resistance in sunflower. *Theor. Appl. Genet.* **2004**, *109*, 92–102. [CrossRef]
18. Román, B.; Torres, A.M.; Rubiales, D.; Cubero, J.I.; Šatović, Z. Mapping of quantitative trait loci controlling broomrape (*Orobanche crenata* Forsk.) resistance in faba bean (*Vicia faba* L.). *Genome* **2002**, *45*, 1057–1063. [CrossRef]
19. Vranceanu, A.V.; Tudor, V.A.; Stoenescu, F.M.; Pirvu, N. Virulence groups of *Orobanche cumana* Wallr. differential hosts and resistance sources and genes in sunflower. In Proceedings of the 9th International Sunflower Conference, Torremolinos, Spain, 8–13 June 1980; pp. 74–82.
20. Flor, H.H. Current Status of the Gene-for-Gene Concept. *Annu. Rev. Phytopathol.* **1971**, *9*, 275–296. [CrossRef]
21. Akhtouch, B.; Munoz-Ruz, J.; Melero-Vara, J.; Fernandez-Martinez, J.; Dominguez, J. Inheritance of resistance to race F of broomrape in sunflower lines of different origins. *Plant Breed.* **2002**, *121*, 266–268. [CrossRef]
22. Fernandez-Martinez, J.; Perez-Vich, B.; Akhtouch, B.; Velasco, L.; Munoz-Ruz, J.; Melero-Vara, J.; Dominguez, J. Registration of Four Sunflower Germplasms Resistant to Race F of Broomrape. *Crop Sci.* **2004**, *44*, 1033–1034. [CrossRef]
23. Perez-Vich, B.; Aktouch, B.; Mateos, A.; Velasco, L.; Jan, C.; Fernández, J.; Dominguez, J.; Fernández-Martínez, J. Dominance relationships for genes conferring resistance to broomrape (*Orobanche cumana* wallr.) in sunflower. *Helia* **2004**, *27*, 183–192. [CrossRef]
24. Akhtouch, B.; del Moral, L.; Leon, A.; Velasco, L.; Martínez, J.M.F.; Pérez-Vich, B. Genetic study of recessive broomrape resistance in sunflower. *Euphytica* **2016**, *209*, 419–428. [CrossRef]
25. Lu, Y.H.; Gagne, G.; Grezes-Besset, B.; Blanchard, P. Integration of a molecular linkage group containing the broomrape resistance gene Or5 into an RFLP map in sunflower. *Genome* **1999**, *42*, 453–456. [CrossRef]
26. Lu, Y.H.; Melero-Vara, J.M.; Garcia-Tejada, J.A.; Blanchard, P. Development of SCAR markers linked to the gene Or5 conferring resistance to broomrape (*Orobanche cumana* Wallr.) in sunflower. *Theor. Appl. Genet.* **2000**, *100*, 625–632. [CrossRef]

27. Tang, S.; Heesacker, A.; Kishore, V.K.; Fernandez, A.; Sadik, E.S.; Cole, G.; Knapp, S.J. Genetic Mapping of the Or5 Gene for Resistance to *Orobanche* Race E in Sunflower. *Crop Sci.* **2003**, *43*, 1021–1028. [CrossRef]
28. Louarn, J.; Boniface, M.-C.; Pouilly, N.; Velasco, L.; Pérez-Vich, B.; Vincourt, P.; Muños, S. Sunflower Resistance to Broomrape (*Orobanche cumana*) Is Controlled by Specific QTLs for Different Parasitism Stages. *Front. Plant Sci.* **2016**, *7*, 590. [CrossRef]
29. Imerovski, I.; Dedić, B.; Cvejic, S.; Miladinović, D.; Jocić, S.; Owens, G.L.; Tubić, N.K.; Rieseberg, L.H. BSA-seq mapping reveals major QTL for broomrape resistance in four sunflower lines. *Mol. Breed.* **2019**, *39*, 41. [CrossRef]
30. Duriez, P.; Vautrin, S.; Auriac, M.C.; Bazerque, J.; Boniface, M.C.; Callot, C.; Carrère, S.; Cauet, S.; Chabaud, M.; Gentou, F.; et al. A receptor-like kinase enhances sunflower resistance to *Orobanche cumana*. *Nat. Plants* **2019**, *5*, 1211–1215. [CrossRef] [PubMed]
31. Goldwasser, Y.; Hershenhorn, J.; Plakhine, D.; Kleifeld, Y.; Rubin, B. Biochemical factors involved in vetch resistance to *Orobanche aegyptiaca*. *Physiol. Mol. Plant Pathol.* **1999**, *54*, 87–96. [CrossRef]
32. Goldwasser, Y.; Plakhine, D.; Kleifeld, Y.; Zamski, E.; Rubin, B. The Differential Susceptibility of Vetch (*Vicia* spp.) to *Orobanche aegyptiaca*: Anatomical Studies. *Ann. Bot.* **2000**, *85*, 257–262. [CrossRef]
33. Yang, C.; Xu, L.; Zhang, N.; Islam, F.; Song, W.; Hu, L.; Liu, D.; Xie, X.; Zhou, W. iTRAQ-based proteomics of sunflower cultivars differing in resistance to parasitic weed *Orobanche cumana*. *Proteomics* **2017**, *17*, 1700009. [CrossRef]
34. Antonova, T.S.; TerBorg, S.J. The role of peroxidase in the resistance of sunflower against *Oronache cumana* in Russia. *Weed Res.* **1996**, *36*, 113–121. [CrossRef]
35. Labrousse, P.; Arnaud, M.C.; Serieys, H.; Bervillé, A.; Thalouarn, P. Several Mechanisms are Involved in Resistance of Helianthus to *Orobanche cumana* Wallr. *Ann. Bot.* **2001**, *88*, 859–868. [CrossRef]
36. Dor, E.; Alperin, B.; Wininger, S.; Ben-Dor, B.; Somvanshi, V.S.; Koltai, H.; Kapulnik, Y.; Hershenhorn, J. Characterization of a novel tomato mutant resistant to the weedy parasites *Orobanche* and *Phelipanche* spp. *Euphytica* **2009**, *171*, 371–380. [CrossRef]
37. Oliveros, J.C. VENNY. An Interactive Tool for Comparing Lists with Venn Diagrams. 2007. Available online: https://bioinfogp.cnb.csic.es/tools/venny/index.html (accessed on 8 February 2017).
38. Simmons, C.R. The physiology and molecular biology of plant 1, 3-β-D-glucanases and 1, 3; 1, 4-β-D-glucanases. *Crit. Rev. Plant Sci.* **1994**, *13*, 325–387.
39. Durrant, W.; Dong, X. Systemic acquired resistance. *Annu. Rev. Phytopathol.* **2004**, *42*, 185–209. [CrossRef] [PubMed]
40. Saboki, E.; Usha, K.; Singh, B. Pathogenesis related (PR) proteins in plant defense mechanism. In *Science against Microbial Pathogens: Communicating Current Research and Technological Advances*; Mendez-Vilas, A., Ed.; Formatex: Badajoz, Spain, 2011; pp. 1043–1047.
41. Van Loon, L. Induced resistance in plants and the role of pathogenesis-related proteins. *Eur. J. Plant Pathol.* **1997**, *103*, 753–765. [CrossRef]
42. Van Loon, L.; Van Strien, E. The families of pathogenesis-related proteins, their activities, and comparative analysis of PR-1 type proteins. *Physiol. Mol. Plant Pathol.* **1999**, *55*, 85–97. [CrossRef]
43. Castillejo, M.; Amiour, N.; Dumas-Gaudot, E.; Rubiales, D.; Jorrín, J.V. A proteomic approach to studying plant response to crenate broomrape (*Orobanche crenata*) in pea (*Pisum sativum*). *Phytochemistry* **2004**, *65*, 1817–1828. [CrossRef]
44. Anderson, J.; Badruzsaufari, E.; Schenk, P.; Manners, J.M.; Desmond, O.J.; Ehlert, C.; MacLean, D.J.; Ebert, P.; Kazan, K. Antagonistic Interaction between Abscisic Acid and Jasmonate-Ethylene Signaling Pathways Modulates Defense Gene Expression and Disease Resistance in Arabidopsis. *Plant Cell* **2004**, *16*, 3460–3479. [CrossRef]
45. Singh, K.B.; Foley, R.C.; Oñate-Sánchez, L. Transcription factors in plant defense and stress responses. *Curr. Opin. Plant Biol.* **2002**, *5*, 430–436. [CrossRef]
46. Kazan, K. Negative regulation of defence and stress genes by EAR-motif-containing repressors. *Trends Plant Sci.* **2006**, *11*, 109–112. [CrossRef] [PubMed]
47. Fujimoto, S.Y.; Ohta, M.; Usui, A.; Shinshi, H.; Ohme-Takagi, M. Arabidopsis ethylene-responsive element binding factors act as transcriptional activators or repressors of GCC box–mediated gene expression. *Plant Cell* **2000**, *12*, 393–404. [PubMed]
48. McGrath, K.; Dombrecht, B.; Manners, J.M.; Schenk, P.; Edgar, C.I.; Maclean, D.J.; Scheible, W.-R.; Udvardi, M.; Kazan, K. Repressor- and Activator-Type Ethylene Response Factors Functioning in Jasmonate Signaling and Disease Resistance Identified via a Genome-Wide Screen of Arabidopsis Transcription Factor Gene Expression. *Plant Physiol.* **2005**, *139*, 949–959. [CrossRef] [PubMed]
49. Liu, S.; Wang, P.; Liu, Y.; Wang, P. Identification of candidate gene for resistance to broomrape (*Orobanche cumana*) in sunflower by BSA-seq. *Oil Crop Sci.* **2020**, *5*. [CrossRef]
50. Parker, C.; Dixon, N. The use of polyethylene bags in the culture and study of *Striga* spp. and other organisms on crop roots. *Ann. Appl. Biol.* **1983**, *103*, 485–488. [CrossRef]
51. Eizenberg, H.; Plakhine, D.; Hershenhorn, J.; Kleifeld, Y.; Rubin, B. Resistance to broomrape (*Orobanche* spp.) in sunflower (*Helianthus annuus* L.) is temperature dependent. *J. Exp. Bot.* **2003**, *54*, 1305–1311. [CrossRef]
52. Hoagland, D.R.; Arnon, D.I. The water-culture method for growing plants without soil. *Circ. Calif. Agric. Exp. Stn.* **1950**, *347*, 32.
53. Ruzin, S.E. *Plant Microtechnique and Microscopy*; Oxford University Press: New York, NY, USA, 1999; Volume 198.
54. Andrews, S. FastQC: A Quality Control Tool for High Throughput Sequence Data. 2010. Available online: http://www.bioinformatics.babraham.ac.uk/projects/fastqc/ (accessed on 29 December 2016).
55. Bolger, A.M.; Lohse, M.; Usadel, B. Trimmomatic: A flexible trimmer for Illumina sequence data. *Bioinformatics* **2014**, *30*, 2114–2120. [CrossRef]

56. Badouin, H.; Gouzy, J.; Grassa, C.; Murat, F.; Staton, S.; Cottret, L.; Lelandais-Brière, C.; Owens, G.L.; Carrère, S.; Mayjonade, B.; et al. The sunflower genome provides insights into oil metabolism, flowering and Asterid evolution. *Nature* **2017**, *546*, 148–152. [CrossRef]
57. Dobin, A.; Davis, C.A.; Schlesinger, F.; Drenkow, J.; Zaleski, C.; Jha, S.; Batut, P.; Chaisson, M.; Gingeras, T. STAR: Ultrafast universal RNA-seq aligner. *Bioinformatics* **2012**, *29*, 15–21. [CrossRef]
58. Li, B.; Dewey, C.N. RSEM: Accurate transcript quantification from RNA-Seq data with or without a reference genome. *BMC Bioinform.* **2011**, *12*, 323. [CrossRef] [PubMed]
59. Love, M.I.; Huber, W.; Anders, S. Moderated estimation of fold change and dispersion for RNA-seq data with DESeq2. *Genome Biol.* **2014**, *15*, 550. [CrossRef] [PubMed]
60. Benjamini, Y.; Hochberg, Y. Controlling the False Discovery Rate: A Practical and Powerful Approach to Multiple Testing. *J. R. Stat. Soc. Ser. B (Methodol.)* **1995**, *57*, 289–300. [CrossRef]
61. Götz, S.; Garcia-Gomez, J.M.; Terol, J.; Williams, T.; Nagaraj, S.H.; Nueda, M.J.; Robles, M.; Talón, M.; Dopazo, J.; Conesa, A. High-throughput functional annotation and data mining with the Blast2GO suite. *Nucleic Acids Res.* **2008**, *36*, 3420–3435. [CrossRef] [PubMed]

Article

The Effect of a Host on the Primary Metabolic Profiling of Cuscuta Campestris' Main Organs, Haustoria, Stem and Flower

Krishna Kumar [1] and Rachel Amir [1,2,*]

1 Migal—Galilee Technology Center, P.O. Box 831, Kiryat Shmona 1101600, Israel; krishnasahni16@gmail.com
2 Tel-Hai College, Kfar Giladi 1220800, Israel
* Correspondence: rachel@migal.org.il; Tel.: +972-4-6953516; Fax: 972-4-6944980

Abstract: *Cuscuta campestris* (dodder) is a stem holoparasitic plant without leaves or roots that parasitizes various types of host plants and causes damage to certain crops worldwide. This study aimed at gaining more knowledge about the effect of the hosts on the parasite's levels of primary metabolites. To this end, metabolic profiling analyses were performed on the parasite's three main organs, haustoria, stem and flowers, which developed on three hosts, *Heliotropium hirsutissimum*, *Polygonum equisetiforme* and *Amaranthus viridis*. The results showed significant differences in the metabolic profiles of C. campestris that developed on the different hosts, suggesting that the parasites rely highly on the host's metabolites. However, changes in the metabolites' contents between the organs that developed on the same host suggest that the parasite can also self-regulate its metabolites. Flowers, for example, have significantly higher levels of most of the amino acids and sugar acids, while haustoria and stem have higher levels of several sugars and polyols. Determination of total soluble proteins and phenolic compounds showed that the same pattern is detected in the organs unrelated to the hosts. This study contributes to our knowledge about the metabolic behavior of this parasite.

Keywords: GC-MS analysis; holoparasitic plant; metabolic changes; parasitic organs

Citation: Kumar, K.; Amir, R. The Effect of a Host on the Primary Metabolic Profiling of Cuscuta Campestris' Main Organs, Haustoria, Stem and Flower. *Plants* **2021**, *10*, 2098.
https://doi.org/10.3390/plants10102098

Academic Editors:
Yaakov Goldwasser, Evgenia Dor and Pablo Castillo

Received: 14 August 2021
Accepted: 28 September 2021
Published: 3 October 2021

Publisher's Note: MDPI stays neutral with regard to jurisdictional claims in published maps and institutional affiliations.

1. Introduction

Cuscuta campestris L., also known as dodder, is one of more than 180 species that belong to the Convolvulaceae family. C. campestris is a stem holoparasitic plant, an extensive climber having filiform stems or vines with twining slender stems lacking true roots and leaves. It has a reduced or absent photosynthesis apparatus, therefore, its development and growth rely on autotropic host plants for at least carbohydrates [1–3]. After proper attachment to the host's vascular system (xylem and phloem) through the haustoria [2], the parasite functions as an active sink, redirecting solutes away from autotrophic sink tissues. Due to this parasitism, C. campestris is considered to be among the most destructive agricultural weeds, significantly reducing the yield and quality of the crops' products [3–6]. It attacks many broad-leaf plants, the most sensitive of which include alfalfa, carrot, tomato, sugar beet, onion, potato and several ornamental plants [3]. C. campestris is native to central North America, but in recent decades, it has spread around the world through seed imports, a process that is still ongoing, and thus has become one of the most problematic weeds worldwide [7].

Despite knowledge accumulated on the parasite's lifecycle, mode of action and agricultural damage, our knowledge about the metabolites absorbed by the parasite and how the host affects its primary metabolic profiling is still mostly unknown [6]. Moreover, there is debate in the literature as to whether most of the metabolites found in holoparasites such as C. *campestris* are taken from the host, or whether the parasite produces most of its metabolites on its own, relying primarily on the sugars transported from its hosts. Some studies suggest that the parasite obtains all its necessary compounds from its hosts, including photosynthetic metabolite, nitrogen, macro/micro minerals, water, phytohormones and other primary and secondary metabolites, as well as RNA and proteins (e.g., [1,2,8,9]). However, it was recently found in the C. campestris genome, that this parasite has genes that can function in the synthesis of primary metabolites, such as all fatty acids and amino acids, as well as co-enzymes and vitamins [10,11]. In addition, several reports suggest that holoparasitic plants can self-regulate their own metabolites since their metabolic profiles differ significantly from their hosts [8,12–14]. Still, numerous questions remain unanswered regarding the independence of the parasite from the metabolisms of its host [6].

Metabolomics has proven to be a powerful method for identifying metabolites whose levels are altered during development and growth conditions, as well as in response to stress [15,16]. Such an analysis could also be proposed for putative biochemical pathways and provide more data on the metabolism and physiological processes [15]. Metabolomics techniques have rarely been applied to parasite plant research and focus mainly on secondary metabolites (e.g., [1,4,17]).

The main goal of this study is to gain more knowledge about the metabolism of *C. campestris* using primary metabolic profiling and to examine how its metabolic profiling is affected by its hosts. To this end, we collected three main organs (haustoria, stem and flowers) from parasitic plants that grew on three different hosts.

2. Results

2.1. Primary Metabolic Profiles Analysis Using Gas Chromatography-Mass Spectrometry (GC-MS) Reveals a Distinct Metabolic Profile in Haustoria, Stem and Flowers with Respects to the Hosts

GC-MS analysis was performed to reveal the levels of primary metabolites in three organs of C. campestris that developed on three different hosts: *Heliotropium hirsutissimum*, *Polygonum equisetiforme* and *Amaranthus viridis*. These hosts come from three families: Boraginaceae, Polygonaceae and Amaranthaceae, respectively. From each of these hosts three organs, haustoria, stem and flowers (Figure 1), were collected. All the samples were collected on the same day in plants that were grown in the same wild field. The analysis enabled us to detect 59 annotated primary metabolites belonging to seven distinct biochemical groups: amino acids (15), polyols (5), tricarboxylic acid cycle (TCA) intermediates (4), organic acids (10), sugar acids (4), fatty acids (4), others (3) and sugars (14). The sugars also included three unannotated sugars (NA1, NA2 and NA3), which the GC-MS identified as sugars but could not indicate the right annotation according to the m/z ratio. The other annotated sugars and metabolites were identified by standards or by the retention index relative to alkane's standards.

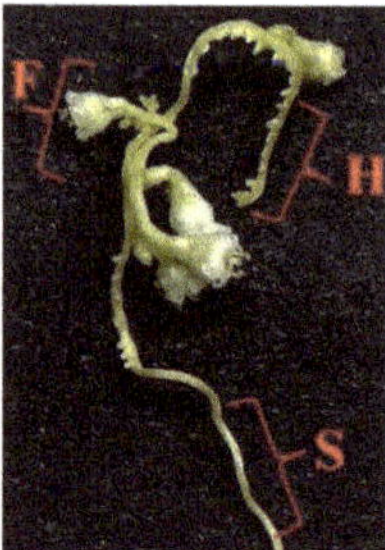

Cuscuta campestris

Figure 1. The organs of *Cuscuta campestris*, where F, H and S indicate flowers, haustoria and stem, respectively.

To obtain more information, the levels of these metabolites in the three organs in each parasite that developed on the three hosts were organized in Table 1. The results showed that flowers from the three hosts have relatively higher levels of the branch chain amino acids, isoleucine and valine, as well as alanine and phenylalanine. This suggests that these amino acids can synthesize and/or accumulate in the flowers of this parasite. High levels of other amino acids are found in one or two hosts, such as proline, which had high levels in the flowers of *H. hirsutissimum*, as well as tyrosine and tryptophan (both belonging to the aromatic amino acids), whose levels were found to be higher in *H. hirsutissimum* and *P. equisetiforme* (Table 1). Flowers also showed significantly higher levels of sugar acids (glucuronic acid, galactonic acid and gluconic acid), as well as sugars (fructose, talose, sugar NA3 and gentobiose). They also had higher levels of benzoic acid, gamma-butyric acid, phosphoric acid and glycerol compared to the two other organs that developed on the same parasite (Table 1). The rest of the metabolites showed a comparatively large variety between the parasite's flowers that developed on the three hosts.

Table 1. The levels of individual primary metabolites in the organs (flowers, haustoria and stem) of *Cuscuta campestris* that developed on three hosts, *Heliotropium hirsutissimum*, *Polygonum equisetiforme* and *Amaranthus viridis* as detected by using GC-MS. The data represent the mean ± SE of three biological replicates from each organ. Values are relative peak areas normalized to the norleucine internal standard. Quantities of soluble amino acids were calculated as nmol/g dry weight. Orange, blue and green colors correspond to metabolites that detected in flower, haustoria and stem. Significance was calculated according to the Tukey Kramer HSD test ($p < 0.05$) and identified by letters. NA, non-annotated sugars; ND, not detected or the values were less than 1.

	Flowers			Haustoria			Stem		
Metabolites	Hh	Pe	Av	Hh	Pe	Av	Hh	Pe	Av
				Sugars					
Sucrose	229 ± 46 [e]	366 ± 16 [cd]	2 ± 0.03 [f]	710 ± 9 [a]	463 ± 17 [bc]	763 ± 40 [a]	535 ± 8 [b]	463 ± 10 [bc]	284 ± 1 [de]
Glucose	373 ± 36 [bc]	473 ± 21 [ab]	325 ± 5 [bc]	183 ± 9 [bc]	253 ± 7 [bc]	30 ± 1 [c]	747 ± 215 [a]	485 ± 17 [ab]	32 ± 2 [c]
Trehalose	38 ± 8 [e]	78 ± 2 [c]	372 ± 6 [d]	159 ± 2 [a]	110 ± 4 [b]	152 ± 7 [a]	113 ± 1 [b]	107 ± 1 [c]	55 ± 1 [de]
Galactose	828 ± 69 [c]	988 ± 36 [b]	794 ± 4 [cd]	614 ± 18 [e]	659 ± 16 [de]	284 ± 12 [f]	1206 ± 12 [a]	1318 ± 30 [a]	168 ± 5 [f]
Talose	54 ± 2 [b]	65 ± 1 [a]	41 ± 0.2 [c]	10 ± 1 [e]	32 ± 1 [d]	3 ± 0.2 [f]	27 ± 1 [d]	42 ± 1 [c]	1 ± 0.07 [f]
Fructose	115 ± 5 [ab]	137 ± 4 [a]	97 ± 1 [bc]	14 ± 1 [e]	54 ± 1 [d]	11 ± 0.5 [e]	48 ± 10 [d]	81 ± 9 [c]	5 ± 0.5 [e]
Glucopyranose	367 ± 14 [a]	255 ± 54 [ab]	18 ± 0.9 [d]	24 ± 1 [d]	18 ± 0.8 [d]	76 ± 3 [d]	238 ± 38 [bc]	122 ± 19 [cd]	9 ± 0.5 [d]
Mannose	2 ± 0.1 [bc]	1 ± 0.02 [bc]	460 ± 2 [a]	4 ± 0 [b]	ND	ND	2 ± 0.02 [bc]	ND	ND
Cellobiose	3 ± 1 [c]	39 ± 2 [a]	7 ± 0.4 [c]	1 ± 0.1 [c]	13 ± 1 [c]	31 ± 3 [ab]	16 ± 7 [bc]	30 ± 2 [ab]	13 ± 0.9 [c]
Gentobiose	11 ± 0.5 [a]	8 ± 1 [b]	2 ± 0.1 [de]	1 ± 0.06 [de]	2 ± 0.01 [d]	1 ± 0.02 [e]	1 ± 0.02 [de]	4 ± 0.6 [c]	1 ± 0.05 [e]
Xylose	11 ± 1 [abc]	11 ± 1 [ab]	5 ± 0.3 [cd]	3 ± 0.09 [de]	7 ± 0.4 [bcd]	1 ± 0.01 [e]	16 ± 2 [a]	13 ± 1 [a]	1 ± 0.02 [e]
Sugar (NA1)	1 ± 0.04 [c]	2 ± 0.3 [c]	ND	5 ± 3 [a]	2 ± 0.05 [ab]	6 ± 0.9 [a]	1 ± 0.01 [c]	2 ± 0.2 [ab]	1 ± 0.01 [c]
Sugar (NA2)	50 ± 1 [c]	123 ± 2 [b]	98 ± 2 [b]	23 ± 1 [cd]	32 ± 0.8 [cd]	6 ± 0.6 [d]	193 ± [a]	134 ± 2 [b]	4 ± 0.8 [d]
Sugar (NA3)	513 ± 4 [a]	358 ± 9 [b]	297 ± 3 [b]	5 ± 0.7 [d]	2 ± 0.01 [d]	13 ± 0.8 [d]	379 ± 6 [ab]	232 ± 23 [c]	1 ± 0.07 [d]
				Sugar acids					
Galactonic acid	119 ± 1.9 [b]	83 ± 1 [c]	148 ± 4 [a]	1 ± 0.02 [d]	2 ± 0.1 [d]	5 ± 0.4 [d]	2 ± 0.4 [d]	7 ± 0.1 [d]	2 ± 0.06 [d]
Glucuronte	15 ± 0.9 [a]	11 ± 0.5 [b]	109 ± 1 [c]	ND	ND	1 ± 0.02 [d]	ND	1 ± 0.04 [d]	ND
Gluconic acid	242 ± 38 [a]	207 ± 5 [a]	40 ± 1 [b]	2 ± 0.01 [b]	ND	16 ± 1 [b]	4 ± 0.6 [b]	18 ± 1 [b]	5 ± 0.5 [b]
Galactaric acid	4 ± 0.4 [cd]	3 ± 0.4 [d]	5 ± 0.8 [b]	2 ± 0.06 [e]	2 ± 0.06 [e]	6 ± 0.7 [a]	1 ± 0.05 [e]	4 ± 0.9 [bc]	2 ± 0.3 [e]
				TCA metabolites					
Malic acid	329 ± 10 [b]	244 ± 5 [c]	150 ± 1 [d]	332 ± 5 [b]	226 ± 5 [c]	381 ± 8 [a]	365 ± 15 [ab]	328 ± 3 [b]	150 ± 1 [d]
Citric acid	264 ± 13 [a]	191 ± 3 [cd]	152 ± 1 [f]	241 ± 2 [ab]	161 ± 2 [ef]	186 ± 5 [de]	219 ± 7 [bc]	230 ± 2 [b]	90 ± 0.9 [g]
Succinic acid	44 ± 2 [abc]	19 ± 0.7 [bc]	77 ± 1 [a]	43 ± 15 [abc]	11 ± 0.8 [c]	13 ± 0.6 [c]	15 ± 1 [c]	61 ± 21 [ab]	6 ± 0.4 [c]
Fumaric acid	73 ± 2 [a]	22 ± 0.7 [cd]	21 ± 0.9 [d]	17 ± 0.9 [de]	8 ± 0.5 [ef]	40 ± 0.4 [b]	76 ± 4 [a]	13 ± 0.3 [f]	30 ± 1 [c]

Table 1. *Cont.*

Metabolites	Flowers Hh	Flowers Pe	Flowers Av	Haustoria Hh	Haustoria Pe	Haustoria Av	Stem Hh	Stem Pe	Stem Av
Organic acids									
Shikimic acid	645 ± 45 a	309 ± 32 bc	224 ± 3 cd	279 ± 13 bc	173 ± 9 de	278 ± 16 bc	548 ± 17 a	380 ± 12 b	123 ± 1 e
Benzoic acid	16 ± 0.8 a	12 ± 0.7 b	9 ± 0.9 c	5 ± 0.4 e	8 ± 0.6 d	3 ± 0.5 f	6 ± 0.6 e	12 ± 0.5 b	1 ± 0.1 g
Pyroglutamate	10 ± 0.7 ab	5 ± 0.6 de	9 ± 0.6 ab	12 ± 1 a	7 ± 0.5 bcd	5 ± 0.4 d	6 ± 0.5 cd	6 ± 1 bcd	1 ± 0.04 e
Nicotinic acid	3 ± 0.2 a	2 ± 0.02 ab	2 ± 0.3 ab	1 ± 0 bc	1 ± 0.01 abc	1 ± 0.01 bc	1 ± 0.06 bc	2 ± 0.02 ab	1 ± 0.06 c
Quinic acid	1 ± 0.01 d	1 ± 0.1 cd	42 ± 1 cd	4 ± 0 a	2 ± 0.3 bcd	4 ± 0.1 a	2 ± 0.2 bc	2 ± 0.3 bcd	1 ± 0.03 cd
Butanoic acid	1 ± 0.2 a	ND	ND	ND	ND	ND	ND	1 ± 0.01 a	ND
Propanoic acid	23 ± 1 a	10 ± 0.3 bcd	12 ± 0.4 b	6 ± 0.5 def	5 ± 0.6 ef	7 ± 0.6 de	7 ± 1 cde	11 ± 1 bc	3 ± 0.02 f
Phosphoric acid	327 ± 4 a	246 ± 1 b	174 ± 3 c	99 ± 6 e	147 ± 5 d	49 ± 0.4 f	110 ± 7 e	241 ± 4 b	25 ± 0.4 g
Erythronic acid	86 ± 6 b	51 ± 1 c	32 ± 0.6 d	142 ± 6 a	148 ± 1 e	19 ± 0.9 de	65 ± 2 c	32 ± 1 d	23 ± 1 de
GABA	57 ± 1 a	22 ± 0.2 c	18 ± 0.9 c	34 ± 1 b	13 ± 0.7 d	6 ± 0.7 e	10 ± 0.9 d	2 ± 0.4 f	1 ± 0.07 f
Polyols									
Mannitol	87 ± 35 bc	135 ± 2 ab	134 ± 1 ab	53 ± 2 cd	173 ± 4 a	3 ± 0.8 d	156 ± 4 a	139 ± 2 ab	4 ± 2 d
Xylitol	26 ± 11 cd	33 ± 1 abcd	30 ± 0.1 b	50 ± 1 a	46 ± 0.9 abc	45 ± 1 abc	43 ± 1 abc	47 ± 2 ab	18 ± 1 d
Inositol	646 ± 24 a	488 ± 12 c	2 ± 0 g	470 ± 5 cd	317 ± 5 e	260 ± 14 e	563 ± 10 b	430 ± 9 d	96 ± 3 f
Galactinol	20 ± 2 c	33 ± 2 bc	76 ± 1 ab	13 ± 1 c	7 ± 0.4 c	61 ± 24 ab	24 ± 1 bc	7 ± 0.6 c	63 ± 1 a
Sorbitol	187 ± 80 cd	686 ± 11 ab	303 ± 2 c	75 ± 3 de	677 ± 16 b	10 ± 1 e	229 ± 5 e	818 ± 9 a	13 ± 1 e
Others									
Ethanolamine	83 ± 8 cd	70 ± 1 cd	56 ± 0 def	88 ± 2 c	157 ± 4 a	63 ± 3 cde	41 ± 6 ef	122 ± 12 b	29 ± 1 f
Glycerol	265 ± 1 a	148 ± 2 b	146 ± 2 b	54 ± 2 e	69 ± 2 de	22 ± 0.9 f	88 ± 3 cd	96 ± 13 c	15 ± 1 f
Lumichrome	ND	1 ± 0.07 bc	ND	ND	2 ± 0.1 a	2 ± 0.03 a	ND	1 ± 0.05 b	ND
Fatty acids									
Hexadecanoate	10 ± 1 a	ND	1 ± 0.1 bc	10 ± 0.07 a	1 ± 0.01 c	8 ± 0.8 a	1 ± 0.02 c	7 ± 3 a	7 ± 0.3 ab
Octadecanoate	2 ± 1 a	ND	ND	1 ± 0.7 a	ND	2 ± 0.07 a	ND	2 ± 1 a	2 ± 0.01 a
Octadecadienoate	3 ± 1 c	ND	ND	7 ± 0.08 ab	1 ± 0.01 d	6 ± 0.7 ab	1 ± 0.01 d	7 ± 0.1 a	5 ± 0.7 bc
Octadecatrienoat	6 ± 0.5 ab	ND	1 ± 0.03 b	7 ± 2 ab	2 ± 2 b	12 ± 0.8 a	1 ± 1 b	8 ± 3 ab	7 ± 0.7 ab
Amino acids									
Alanine	1210 ± 44 a	888 ± 18 b	965 ± 74 b	521 ± 87 c	665 ± 30 c	516 ± 3 c	695 ± 24 c	665 ± 39 c	655 ± 15 c
Valine	3782 ± 232 b	5324 ± 205 a	3347 ± 48 b	1124 ± 47 d	1927 ± 81 c	851 ± 72 d	1183 ± 35 cd	1601 ± 120 cd	1307 ± 95 cd
Serine	5200 ± 188 b	5401 ± 107 b	9591 ± 1504 a	2890 ± 173 c	4495 ± 265 bc	3031 ± 178 bc	3194 ± 71 bc	3704 ± 722 bc	2893 ± 231 bc
Leucine	2411 ± 70 a	2113 ± 26 abc	2180 ± 103 ab	768 ± 12 abc	1037 ± 6 cd	659 ± 36 d	973 ± 9 d	1130 ± 19b cd	1055 ± 39 cd
Threonine	627 ± 19 ab	737 ± 12 a	590 ± 167 ab	551 ± 8 b	713 ± 27 ab	439 ± 24 b	707 ± 5 ab	729 ± 40 a	621 ± 24 ab
Isoleucine	1199 ± 37 a	1374 ± 24 a	1290 ± 57 a	566 ± 10 bc	794 ± 13 b	600 ± 29 bc	601 ± 12 c	751 ± 18 bc	767 ± 31 b
Proline	1026 ± 58 a	198 ± 15 cd	279 ± 22 cd	639 ± 13 b	177 ± 41 cd	127 ± 18 d	328 ± 35 c	153 ± 49 cd	179 ± 11 cd
Glycine	782 ± 53 a	921 ± 166 a	637 ± 74 a	311 ± 47 a	614 ± 38 a	544 ± 30 a	440 ± 81 a	369 ± 80 a	510 ± 40 a
Homoserine	327 ± 21 a	252 ± 8 abc	267 ± 28 abcd	176 ± 2 cd	238 ± 29 ab	158 ± 14 d	223 ± 1 abcd	204 ± 4 bcd	221 ± 13 abcd
Methionine	270 ± 9 ab	396 ± 4 a	367 ± 19 a	215 ± 21 b	234 ± 47 b	162 ± 5 b	267 ± 5 ab	197 ± 13 b	270 ± 7 ab
Phenylalanine	562 ± 16 a	499 ± 4 ab	429 ± 26 abc	241 ± 1 bcd	300 ± 2 cd	193 ± 1 d	281 ± 4 cd	327 ± 5 bcd	327 ± 18 bcd
Glutamate	20294 ± 110 a	7387 ± 243 cd	14981 ± 789 ab	15377 ± 295 bcd	4623 ± 1194 d	11007 ± 1358 bc	14600 ± 865 abc	12221 ± 1286 ab	20197 ± 402 a
Glutamine	2130 ± 152 d	7629 ± 895 cd	5743 ± 901 cd	41281 ± 3025 a	8196 ± 661 cd	15172 ± 1118 b	35770 ± 1484 a	8322 ± 614 bc	13122 ± 1049 bc
Tyrosine	26612 ± 101 a	2400 ± 197 b	15255 ± 181 ab	3395 ± 463 ab	1318 ± 84 b	5891 ± 573 ab	7072 ± 539 ab	7025 ± 224 ab	9318 ± 943 ab
Tryptophan	4109 ± 113 a	2628 ± 87 b	1047 ± 214 c	765 ± 193 c	1251 ± 27 c	666 ± 70 c	727 ± 78 c	833 ± 124 c	1086 ± 252 c

The haustoria from the parasite that developed on the three hosts tended to have high levels of sucrose, trehalose, sugar NA1, quinic acid, xylitol, ethanolamine, lumichrome and the amino acid, glutamine. The stems of C. campestris that grew on *H. hirsutissimum* and *P. equisetiforme* showed significantly higher levels of xylose, galactose, sugar NA2 and malic acid (Table 1).

Taken together, the results indicate that the hosts significantly affect the levels of metabolites in the three organs. This effect is more pronounced in the stems and haustoria, and less in the flowers, which showed a greater number of common metabolites that arose in this organ relative to the two other organs.

The results also showed large differences in the accumulation of the metabolites between the three organs of C. *campestris* that developed on the same host. To test the general effect of the host on the accumulation of metabolites on the three different organs of C. *campestris*, the metabolic profiles of the haustoria, stem and flowers were plotted onto a principal component analysis (PCA) that mathematically quantifies the distance between variables and expresses the original data by principal components in the plotting space [18] (Figure 2). Variances were explained by the two components, PC1, which was responsible for 66.2%, and PC2, which gave a value of 18.3% of the variance. The observation that each of the organs from the three hosts, haustoria, stem and flowers, do not cluster together further strengthens the impression that each of the three hosts significantly affects the metabolic profiling of each of the three organs in different ways. The finding that the organs of *P. equisetiforme* are relatively closer to each other compared to the other two hosts suggests that this host is more affected the metabolites in the organs. The haustoria that developed on this host was relatively far from those developed on the two other hosts (Figure 2).

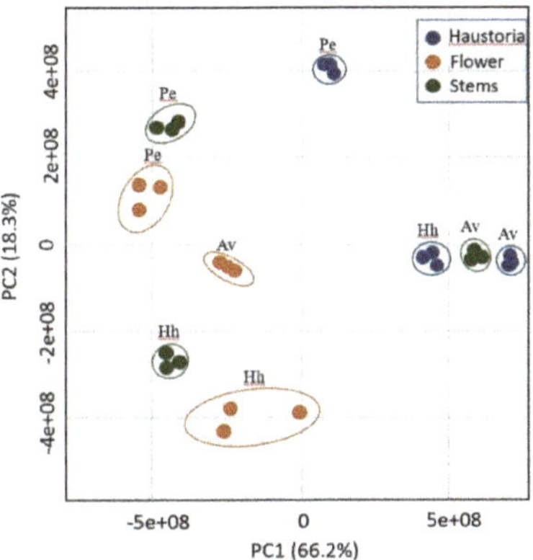

Figure 2. Principal component analyses (PCA) applied to *Cuscuta campestris* organs that developed on three hosts [*Heliotropium hirsutissimum* (Hh), *Polygonum equisetiforme* (Pe), *Amaranthus viridis* (Av)], according to their entire primary metabolome set of 59 metabolites. The data points are displayed as projections onto the two primary axes (eigenvectors). Variances explained by the first two components (PC1 and PC2) appear in parentheses.

In addition, the results show that the host's metabolites mainly affected the stem's metabolites, since the distance between the clusters of this organ that developed on the three hosts was relatively far compared to the other two organs (flowers and haustoria). The clusters representing the stems, and to a lesser extent also the haustoria, were scattered on both the transverse and longitudinal axes (PC1 and PC2), while the flower components obtained from the three hosts were mainly affected by PC2 (Figure 2). This suggests that the differences between the flowers are smaller relative to the other two organs. The clusters that represent the organs developed on the parasite that infected the *A. viridis* that was influenced by PC1.

To gain more knowledge about the effect of the host on the metabolites that accumulate in each organ, biplot analyses were performed using an R-based software (Figure 3). When the flowers of *C. campestris* collected from the three hosts were plotted together, it can be shown that sorbitol, shikimic acid, galactose, sucrose, glucopyranose and glucose are the metabolites that mostly affected the variance (Figure 3A). The levels of sorbitol, inositol, glucose, galactose and eryhronic acid affected the haustoria (Figure 3B). The variance of the stems was affected by sorbitol, galactose, shikimic acid, glucopyranose, sucrose, glucose and sugar NA3 (Figure 3C). Taken together, sorbitol, inositol, galactose and glucose are the main metabolites that contribute to variance when each of the organs was examined separately (Figure 3).

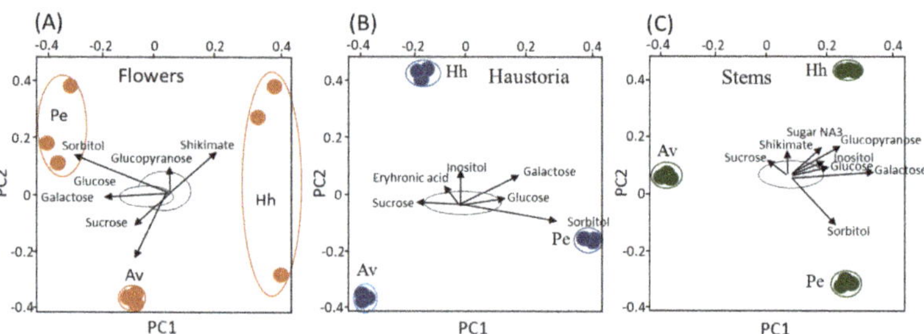

Figure 3. Biplot analysis applied to each of three *Cuscuta campestris* organs (haustoria, stem and flowers) that developed on each of three different host plants (**A–C**). The area marked with a black line is the area where most of the metabolites are located. Only the metabolites that have exceeded the boundaries of the area are marked with arrows. The data represent three replicates for each organ. The host plants are *Heliotropium hirsutissimum* (Hh), *Polygonum equisetiforme* (Pe) and *Amaranthus viridis* (Av).

We also performed a biplot analysis on the three organs of *C. campestris* that developed on each of the hosts. When the organs were collected from the parasite grown on the *H. hirsutissimum* plot together, it was shown that metabolites such as gluconic acid, glycerol, glucopyranose, sugar NA3, galactonic acid, fructose and shikimate contributed mostly to the variance of the flowers, while those contributing to the variance of the stems were galactose, glucose, sugar NA2 and malic acid (Figure 4A). Sucrose and trehalose contributed to the haustoria. The results showed that the metabolites that contributed to the variance were distributed in both PC1 and PC2. When the same analysis was made for the parasite that developed on *P. equisetiforme*, it was defined that as detected in the flowers of *H. hirsutissimum*, gluconic acid, glucopyranose, sugar NA3 and galactonic acid contributed to the flowers in addition to inositol. The levels of galactose, glucose and shikimate mainly contribute to the variance of the stems that developed on this host (Figure 4B). The metabolites that mostly affected the variance of the flowers of *C. campestris* collected on *A. viridis* were galactose, glucose, inositol, sugar NA3 and glucopyranose (Figure 4C). Galactinol contributes to the variance of the stems. Together, the metabolites on this host mostly contribute to PC1, while those of *H. hirsutissimum* contribute to PC1 and PC2, and those of *P. equisetiforme* were in an intermediate state between the two other hosts (A–C). Most of the metabolites in the parasites that developed on the three hosts contributed to the variance of the flowers and stems, but much less to the haustoria (Figure 4A–C).

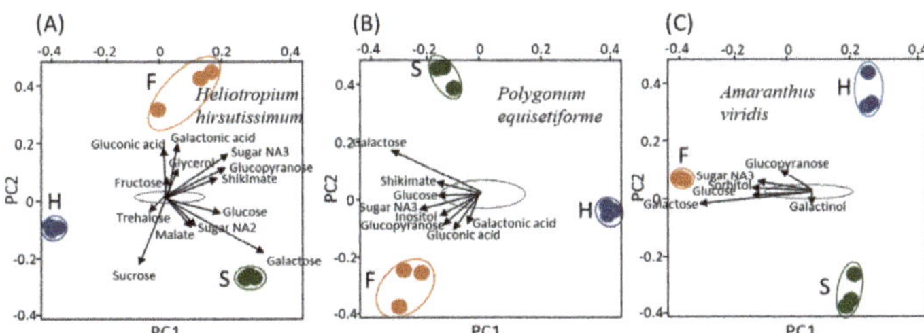

Figure 4. Biplot analysis applied to each of the three organs (haustoria, stem and flowers) of *Cuscuta campestris* that developed on the same host (**A–C**). The area marked with a black line is the area where most of the metabolites are located. Only the metabolites that have exceeded the boundaries of the area are marked with arrows. The data represent three replicates for each organ. The host plants are *Heliotropium hirsutissimum*, *Polygonum equisetiforme* and *Amaranthus viridis*; F, H and S indicate flowers, haustoria and stem, respectively.

To further verify the effect of the host on the accumulation of metabolites in the three organs and to determine general trends, a heat-map analysis was performed (Figure 5). As indicated in the PCA analysis, the results show that the metabolic profile of the flowers that were collected from the parasite that grew on the different hosts is relatively similar (Figure 5). Compared to the stem and haustoria, the flowers accumulate high levels of several metabolites, which are comprised of several amino acids (serine, valine, isoleucine, leucine, methionine, glycine, phenylalanine, gamma-butyric acid, alanine, homoserine), sugars (glucopyranose, talose, fructose, gentobiose, sugars NA3), sugar acids (gluconic acid, galactorinc acid, glucuronic acid), the two polyols (inositol, glycerol), as well as succinic acid, propanoic acid and nicotinic acid (Figure 5). The stems and haustoria, however, tend to have relatively high levels of other metabolites compared to the flowers such as other sugars (sucrose, trehalose, sugar NA1), organic acids (quinic acid, malic acid) and polyols (xylitol).

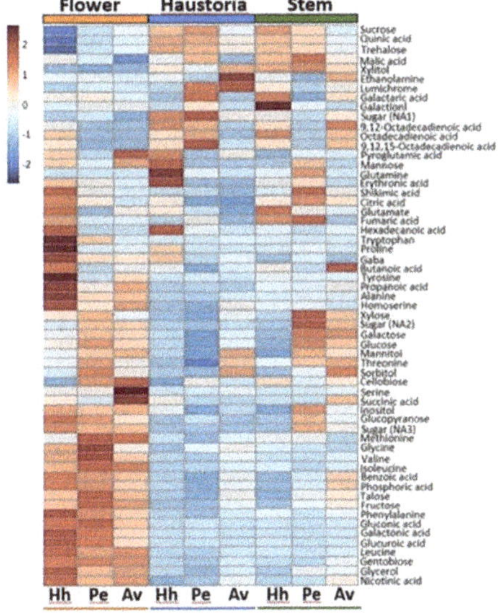

254

Figure 5. Heat-map of these 59 primary metabolites detected by GC-MS. The data represent three replicates. The host plants are *Heliotropium hirsutissimum* (Hh), *Polygonum equisetiforme* (Pe) and *Amaranthus viridis* (Av).

2.2. Determination of Total Soluble Protein and Total Phenolic Compounds

The higher levels of most of the free amino acids in the flowers irrespective of the hosts suggested that the flowers have higher levels of proteins. To examine this assumption, a Bradford analysis was performed on the soluble protein fractions that were extracted from the different organs. The analyses revealed that the protein content indeed tends to be highest in the flowers, followed by the haustoria, whereas in the stem it is significantly lowered irrespective of the host plants (Figure 6A). In addition, the relatively high levels of the aromatic amino acids in the organs of *C. campestris* suggest that they can influence the synthesis of phenolic compounds. To verify this assumption, the levels of total soluble phenolic compounds were examined in the organs. The results demonstrated that the total phenolic content was highest in the haustoria obtained from *C. campestris* that grew on the three hosts, followed by the stem, and the lowest amount accumulated was detected in the flower (Figure 6B).

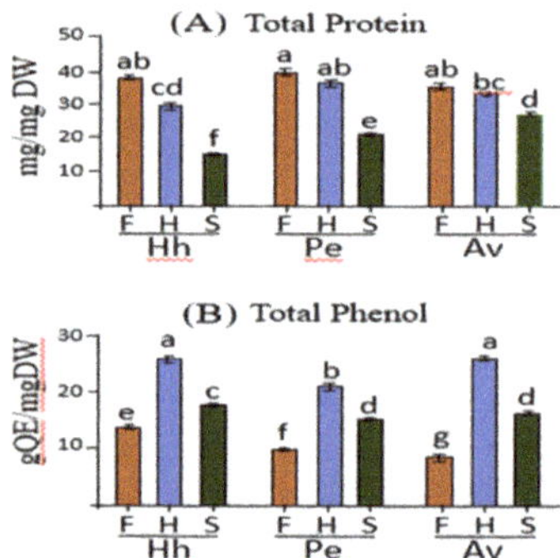

Figure 6. The total protein (**A**) and total phenol content (**B**) in the three organs of *Cuscuta campestris*, flowers (F), haustoria (H) and stem (S), that developed on the three host plants, *Heliotropium hirsutissimum* (Hh), *Polygonum equisetiforme* (Pe) and *Amaranthus viridis* (Av). (**A**) The total protein contents in the albumin fraction as measured using the Bradford assay; (**B**) The total phenol contents represented as mg quercetin equivalents (QE) per mg of dry weight (DW). All data shown are means $\pm$ SE of three replicates for each organ. The significance was calculated according to the Tukey-Kramer HSD test ($p < 0.05$) and is identified by different small letters.

3. Discussion

The main aim of the current study was to determine the effect of three different hosts on the primary metabolic profile of three main organs of *C. campestris* (haustoria, stem and flowers). The leading assumption was that *C. campestris*, similar to all other holoparasite plants such as different species of *Phelipanche*, *Orobanchae*, *Conopholis* and *Epifagus* [4], relies on their hosts for carbohydrates, minerals and water to complete their life cycle [4,17,19]. Still, there is little knowledge about the question of how much of the *C. campestris*' metabolism relies on its hosts for the other primary metabolites. We assume that if the parasite takes mostly carbohydrates from its hosts and relies on its genes to control the synthesis and accumulation of other metabolites, its metabolic profiling in each of the three organs would be relatively similar when it grows on different hosts. However, if its metabolic profiling relies mainly on its hosts' metabolites, its metabolic profiling would differ significantly.

The results of this study showed that the organs of *C. campestris* that developed on the three hosts have different levels of primary metabolites (Figures 2–4; Table 1). This suggests that the host significantly affects the metabolic profiling of *C. campestris* and the parasite strongly relies on the host's metabolites. The results suggest that in addition to photosynthesis products, the parasite absorbs many other primary metabolites, some of which accumulated and others were catabolized to other metabolites.

A similar assumption that the host significantly affects metabolites levels in the parasite was also proposed for another holoparasite, *Phelipanche aegyptiaca* [8]. The hosts were also found to affect the morphological parameters of *C. campestris* that developed on three different hosts since the dry weight, size of stem length and the number of attachment sites of the parasite differed significantly between the three hosts [20]. Despite these studies, dependence on the host for different metabolites might be less necessary at the beginning of the establishment of the parasite; at least one study examined the levels of primary metabolites using GC-MS in *C. japonica* seedlings (7-day-old), and after eight days of parasitization on *Momordica charantia* as a host. It was shown that the levels of only laminaribiose (disaccharide of glucose) increased significantly in the parasite, while the levels of the other detected metabolites were not significantly altered [5].

As shown in the PCA and heatmap analyses, the stems were significantly affected by the host, the haustoria were less affected and the flowers the reproductive organ had relatively the most conserved profile. A similar observation was also derived from the analysis of five different organs of *P. aegyptiaca* that collected in the same mature stage, showing significant differences in the metabolites' levels between the vegetative and reproductive organs [21]. The main metabolites that accumulated in flowers were the soluble amino acids and sugar acids (except for galactaric acid) (Table 1). The high levels of free amino acids in the flowers are in accordance with the high level of total proteins in this organ. This is similar to *P. aegyptiaca*, whose flowers have high levels of sugars, amino acids and total proteins compared to the other organs [21]. The high level of metabolites in the flowers might affect the levels of metabolites in the seeds. However, it was previously shown in *C. campestris* that the seeds had lower levels of carbohydrates and proteins compared to the stems [17]. The high levels of aromatic amino acids in flowers do not link with the observation that this organ has the lowest total phenol content. This finding is in contrast with the positive link between aromatic amino acids and total phenols in the flowers and flower buds of *P. aegyptiaca* [21]. Even though the finding that the *C. campestris'* flowers have relatively low content of total phenols (Figure 5), measuring total phenols in the seeds and shoots of this parasitic plants in another study showed that the levels in seeds are significantly higher than in shoots, which was stable in two dodders irrespective of their different hosts [22], suggesting that seeds can produce or accumulates phenols.

Similar to flowers that show accumulations of certain metabolites, haustoria from all hosts had the highest levels of total phenols compared with the stem and flowers (Figure 5). Since the levels of the aromatic amino acids were relatively low in this organ (Table 1), it raises the possibility that the phenols were derived from the hosts. It is well known that damaged plant tissue accumulates phenols as part of the defense response [23]. Indeed, a previous report suggested that the parasitism of *C. campestris* on host plants induces the synthesis of phenolic compounds in the host [22,24,25]. The finding that the level of phenols in the haustoria differs between plants grown on different hosts also suggests that the phenols come from the host. However, to further verify this assumption, measurements of the levels of phenols in the hosts and flux analysis should be performed.

Do the parasitic plants transport only the metabolites required for their growth? The answer is yet unknown, but it was previously suggested that *C. campestris* and *C. japonica* had no selective absorption of specific compounds from the host [17,26]. Furthermore, an additional study detected secondary metabolites in *C. reflexa* grown on two different plant species, showing that specific compounds that are synthesized in each host were present in the parasite [27]. In any case, these secondary metabolites reflect some of the metabolites that synthesized in their hosts [27]. Moreover, detecting different flavonoids in *C. reflexa* plants growing on five different hosts showed significant differences in the contents of these flavonoids, which reflected their hosts [28]. This variation in phytochemicals present in *C. reflexa* confirms that chemical constituents of the parasite depend on the nature of host and that no selection in the transformed compounds occurs. Yet, it has been suggested that some metabolites transported to dodders could be further metabolized, and indeed studies have shown that in addition to metabolites belonging to the host plant, others were metabolized by Cuscuta [19,29].

Overall, the results of this study have shown that: (i) the levels of primary metabolites of *C. campestris'* organs were affected by their hosts; and (ii) the metabolic profile of at least the flowers is also dictated by the needs of this organ. This latter point is supported by the observation that flowers and also slightly the stems and haustoria have some metabolites that characterized their metabolic profiling, independent of the host. Moreover, the finding that the three organs that developed on the same host showed different profiles suggests that each of the organs has the ability to alter its metabolic profiling by expressing specific sets of genes.

This metabolic study is a first step in understanding the ability of the parasite to accumulate and/or produce its own metabolites. Future studies should focus on assaying the gene expression profile of the parasite and study the enzyme activities, or other proteins involved in the synthesis and accumulation of these metabolites, in order to reveal the biochemical pathways and their regulation in the parasite. Feeding analyses using radiolabeled compounds are also required to reveal the flux from the host towards the parasite, as well as metabolic profiles analyses of the hosts.

4. Materials and Methods

4.1. Plant Material and Sample Collection

Three different organs, haustoria, stem and flowers of C. campestris plant, were collected from three hosts, *Heliotropium hirsutissimum*, *Polygonum equisetiforme* and *Amaranthus viridis*. All the organs were collected on the same day from the same wild field in Kibbutz Dan (northern Israel) (Figure 1). The parasite samples were separated, frozen in liquid nitrogen and lyophilized. The lyophilized organs were then ground to fine powder by mortar and pestle.

4.2. Extraction and Analysis of Primary Metabolites Using GC-MS

Primary metabolites were extracted from 20 mg dried weight of the haustoria, stem and flowers. The samples were homogenized using a Restch MM 301 homogenizer in 1000 μL of methanol/chloroform/double distill water (DDW) (2.5:1:1) at 4 °C. Norleucine (4.6 μL of 2 mg per ml) was added as an internal standard. After short vortex and 10 min of centrifugation at 20,000 g at 4 °C, 1,000 μL the supernatant was collected to a new tube. The lower phase kept for fatty acid analysis. In this case, 300 μL distil DDW and 300 μL chloroform were added. The samples were vortexed for 1 min and settled for 5 min at room temperature, followed by 10 min of centrifugation at 20,000 g at 4 °C. In this case, 300 μL from the upper phase were dried using spcadvac. The dried samples were then dissolved and treated for 2 h at 37 °C with 40 μL 20 mg/mL methoxyamine hydrochloride in pyridine, followed by derivatization for 30 min in N-methyl-N(trimethylsilyl)-trifluoro-acetamide at 37 °C with vigorous shaking. Sample volumes of 1 μL were injected into a GC- 414 MS system. The single-ion mass method was used for soluble amino acid determination with the BP5MS capillary column (SGE-gc; 30 m, 0.25 mm i.d. and 0.25 mm thickness), while the total-ion-count method was used for metabolic profiling and separation using the VF-5ms capillary column (Agilent; 30 m + 10 m EZ-guard, 0.25 mm i.d. and 0.25 mm thicknesses). All analyses were carried out on a GC-MS system (Agilent 7890A) coupled with a mass selective detector (Agilent 5975c) and a Gerstel multipurpose sampler MPS2 [30]. Peak finding, peak integration and retention time correction were performed using the Agilent GC/MSD Productivity ChemStation package (www.agilent.com, last accessed on 25 September 2021). Peak areas were normalized to integral standard (norleucine) signal. For amino acids analysis, amino acid standards of 200, 100, 50, 25 and 5 μM were injected to establish quantification curves, and the amounts of amino acids were calculated accordingly [21].

For fatty acids determination, 200 μL of the lower phase containing fatty acid were then transferred to a new tube and dried with nitrogen gas, followed by the addition of

300 µL of methanol with 2% H2SO4. After vortexed the tubes incubated at 85 °C for one hour under shaking conditions. Then, 300 µL of DDW and 300 µL hexane were added, the blend was mixed well and centrifuged for 20 min at 20,000g. Approximately 150 µL aqueous phase was transferred to the GC-MS tube and analyzed by GC-MS. Heptadecanoic acid was used as an internal standard.

The annotations of the metabolites were made using standards or the retention index relative to alkanes standards. The corresponding mass spectra and retention time indices were compared with standard substances and commercially available electron mass spectrum libraries available from the National Institute of Standards and Technology and Max Planck Institute for Plant Physiology, Golm, Germany (http://www.mpimpgolm.mpg.de/mms-library/, last accessed on 25 September 2021).

4.3. Total Soluble Protein Determination

For total soluble protein determination, 5 mg dried weight of the haustoria, stem and flowers were grounded in 200 µL buffer phosphate pH=7.8 with a protease inhibitor cocktail (Sigma Aldrich). After two centrifugation cycles (20,800 g for 5 min), total protein was determined using a Bradford reagent (Bio-Rad) in three sample concentrations. Bovine serum albumin was used as a standard. Total phenolic compounds content determination For total phenolic compounds content determination, 5 mg dried weight of the haustoria, stem and flowers were ground in 0.5 mL DDW, the colorimetric method that modified the Ben Nasr method for small volumes [31] was used. Six µL of the extraction sample was loaded on a 96 well ELISA plate. To each well, 50 µL of 10% Folin-Ciocalteu reagent and 40 µL of 7.5% (*w/v*) Na_2CO_3 were added. The plate was incubated for 40 min at 37 °C and then read at 765 nm. A standard curve was created using quercetin.

4.4. Statistical Analyses

For the metabolites study, three biological replicate samples were taken of each organ (haustoria, stem and flowers). The data represent the mean of three independent replications for the metabolites and five for the phenol and total protein. Statistical significance was evaluated using JMP software version 8.0 (SAS Institute Inc., Cary, NC). Significant differences between treatments were calculated according to the Tukey-Kramer HSD test (p < 0.05). Principal component analysis (PCA) and a heatmap of GC-MS data were conducted using the MetaboAnalyst 3.0 comprehensive tool (http://metaboanalyst.ca/, last accessed on 25 September 2021; [18] with auto scaling (mean-centered and divided by the standard deviation of each variable) manipulations. Graphs were compiled using GraphPad Prism 5.01 scientific software (http://www.graphpad.com, last accessed on 25 September 2021).

Author Contributions: K.K. and R.A. conducted the study; K.K. carried out the experiments and the analysis; R.A. and K.K. wrote the manuscript. All authors have read and agreed to the published version of the manuscript.

Funding: This work was supported by a grant from the Israeli Ministry of Agriculture and Rural Development Chief Scientist Office, Israel; grant no. 21-01-0036.

Institutional Review Board Statement: Not applicable.

Informed Consent Statement: Not applicable.

Data Availability Statement: Not applicable.

Acknowledgments: We thank Yael Hacham for the critical reading of the manuscript and to Janet Covaliu for her assist with English editing of the manuscript.

Conflicts of Interest: The authors declare no conflict of interest.

References

1. Smith, J.D.; Mescher, M.C.; De Moraes, C.M. Implications of bioactive solute transfer from hosts to parasitic plants. *Curr. Opin. Plant. Biol.* **2013**, *16*, 464–472. [CrossRef]
2. Kim, G.; Westwood, J.H. Macromolecule exchange in Cuscuta-host plant interactions. *Curr. Opin. Plant. Biol.* **2015**, *26*, 20–25. [CrossRef]
3. Saric-Krsmanovic, M.; Vrbnicanin, S. Field dodder-how to control it? *Pestic. I Fitomed.* **2015**, *30*, 137–145. [CrossRef]
4. Ahmad, A.; Tandon, S.; Xuan, T.D.; Nooreen, Z. A Review on Phytoconstituents and Biological activities of Cuscuta species. *Biomed. Pharm.* **2017**, *92*, 772–795. [CrossRef]
5. Furuhashi, T.; Nakamura, T.; Iwase, K. Analysis of metabolites in stem parasitic plant interactions: Interaction of Cuscuta–Momordica versus Cassytha–Ipomoea. *Plants* **2016**, *5*, 43. [CrossRef] [PubMed]
6. Flores-Sánchez, I.J.; Garza-Ortiz, A. Is There a Secondary/Specialized Metabolism in the Genus Cuscuta and which Is the Role of the Host Plant? *Phytochem Rev.* **2019**, *18*, 1299–1335. [CrossRef]
7. Yu, H.; Liu, J.; He, W.M.; Miao, S.L.; Dong, M. Cuscuta australis restrains three exotic invasive plants and benefits native species. *Biol Invasions* **2011**, *13*, 747–756. [CrossRef]
8. Clermont, K.; Wang, Y.; Liu, S.; Yang, Z.; dePamphilis, C.W.; Yoder, J.I. Comparative Metabolomics of Early Development of the Parasitic Plants Phelipanche aegyptiaca and Triphysaria versicolor. *Metabolites* **2019**, *9*, 114. [CrossRef]
9. Nickrent, D.L.; Musselman, L.J. Introduction to Parasitic Flowering Plants. *Plant Health Instr.* **2004**, *13*, 300–315. [CrossRef]
10. Vogel, A.; Schwacke, R.; Denton, A.K.; Usadel, B.; Hollmann, J.; Fischer, K.; Bolger, A.; Schmidt, M.H.W.; Bolger, M.E.; Gundlach, H.; et al. Footprints of parasitism in the genome of the parasitic flowering plant Cuscuta campestris. *Nat. Commun.* **2018**, *9*, 2515. [CrossRef]
11. Clarke, C.R.; Timko, M.P.; Yoder, J.I.; Axtell, M.J.; Westwood, J.H. Molecular Dialog between Parasitic Plants and Their Hosts. *Annu. Rev. Phytopathol.* **2019**, *57*, 279–299. [CrossRef] [PubMed]
12. Hacham, Y.; Hershenhorn, J.; Dor, E.; Amir, R. Primary metabolic profiling of Egyptian broomrape (Phelipanche aegyptiaca) compared to its host tomato roots. *J. Plant Physiol.* **2016**, *205*, 11–19. [CrossRef] [PubMed]
13. Nandula, V.K.; Foster, J.G.; Foy, C.L. Impact of Egyptian broomrape (*Orobanche aegyptiaca* (Pers.) parasitism on amino acid composition of carrot (Daucus carota L.). *J. Agric. Food Chem.* **2000**, *48*, 3930–3934. [CrossRef]
14. Wakabayashi, T.; Joseph, B.; Yasumoto, S.; Akashi, T.; Aoki, T.; Harada, K.; Muranaka, S.; Bamba, T.; Fukusaki, E.; Takeuchi, Y.; et al. Planteose as a storage carbohydrate required for early stage of germination of Orobanche minor and its metabolism as a possible target for selective control. *J. Exp. Bot.* **2015**, *66*, 3085–3097. [CrossRef]
15. Obata, T.; Fernie, A.R. The use of metabolomics to dissect plant responses to abiotic stresses. *Cell Mol. Life Sci.* **2012**, *69*, 3225–3243. [CrossRef] [PubMed]
16. Nakabayashi, R.; Saito, K. Integrated metabolomics for abiotic stress responses in plants. *Curr. Opin. Plant. Biol.* **2015**, *24*, 10–16. [CrossRef]
17. Al-Gburi, B.K.H.; Al-Sahaf, F.H.; Al-fadhal, F.A.; Del Monte, J.P. Detection of phytochemical compounds and pigments in seeds and shoots of Cuscuta campestris parasitizing on eggplant. *Physiol. Mol. Biol. Plants* **2019**, *25*, 253–261. [CrossRef] [PubMed]
18. Xia, J.; Sinelnikov, I.V.; Han, B.; Wishart, D.S. MetaboAnalyst 3.0-making metabolomics more meaningful. *Nucleic Acids Res.* **2015**, *43*, W251–W257. [CrossRef]
19. Kaiser, B.; Vogg, G.; Fürst, U.B.; Albert, M. Parasitic plants of the genus Cuscuta and their interaction with susceptible and resistant host plants. *Front. Plant. Sci* **2015**, *6*, 45. [CrossRef]
20. Koch, A.M.; Binder, C.; Sanders, I.R. Does the generalist parasitic plant Cuscuta campestris selectively forage in heterogeneous plant communities? *New Phytol.* **2004**, *162*, 147–155. [CrossRef]
21. Nativ, N.; Hacham, Y.; Hershenhorn, J.; Dor, E.; Amir, R. Metabolic investigation of Phelipanche aegyptiaca reveals significant changes during developmental stages and in its different organs. *Front. Plant. Sci.* **2017**, *8*, 491. [CrossRef]
22. Löffler, C.; Sahm, A.; Wray, V.; Czygan, F.C.; Proksch, P. Soluble phenolic constituents from Cuscuta reflexa and Cuscuta platyloba. *Biochem. Syst. Ecol.* **1995**, *23*, 121–128. [CrossRef]
23. Isah, T. Stress and defense responses in plant secondary metabolites production. *Biol. Res.* **2019**, *52*, 39. [CrossRef] [PubMed]
24. Özen, H.Ç.; Surmuş Asan, H. The effect of *Cuscuta babylonica* Aucher (Cuscuta) parasitism on the phenolic contents of Carthamus glaucus Bieb.subsp. glaucus. *J. Inst. Sci. Technol.* **2016**, *4*, 31. [CrossRef]
25. Furuhashi, T.; Fragner, L.; Furuhashi, K.; Valledor, L.; Sun, X.; Weckwerth, W. Metabolite changes with induction of Cuscuta haustorium and translocation from host plants. *J. Plant. Interact.* **2012**, *7*, 84–93. [CrossRef]
26. Anis, E.; Ullah, N.; Mustafa, G.; Malik, A.; Alza, N.; Badar, Y. Phytochemical studies on Cuscuta reflexa. *J. Nat. Prod.* **1999**, *5*, 124–126.
27. Bais, N.; Kakkar, A. Phytochemical Analysis of Methanolic Extract of Cuscuta reflexa Grown on Cassia fistula and Ficus benghalensis by GC-MS. *Int. J. Pharm. Sci. Rev. Res.* **2014**, *25*, 33–36.
28. Tanruean, K.; Kaewnarin, K.; Suwannarach, N.; Lumyong, S. Comparative evaluation of phytochemicals, and antidiabetic and antioxidant activities of Cuscuta reflexa grown on different hosts in Northern Thailand. *Nat. Prod. Commun.* **2017**, *12*, 51–54. [CrossRef]
29. Hibberd, J.M.; Jeschke, W.D. Solute flux into parasitic plants. *J. Exp. Bot.* **2001**, *52*, 2043–2049. [CrossRef]

30. Cohen, H.; Israeli, H.; Matityahu, I.; Amir, R. Seed-specific expression of a feedback-insensitive form of CYSTATHIONINE-SYNTHASE in arabidopsis stimulates metabolic and transcriptomic responses associated with desiccation stress. *Plant. Physiol.* **2014**, *166*, 1575–1592. [CrossRef]
31. Ben Nasr, C.; Ayed, N.; Metche, M. Quantitative determination of the polyphenolic content of pomegranate peel. *Z. Lebensm. Unters. Forsch.* **1996**, *203*, 374–378. [CrossRef] [PubMed]

Opinion

A Personal History in Parasitic Weeds and Their Control

Chris Parker

Independent Researcher, 6 Royal York Crescent, Bristol BS8 4JZ, UK; chrisparker5@compuserve.com

Abstract: This invited paper summarises a career in which I became increasingly involved in research and related activities on *Striga* and other parasitic weeds. It also presents a personal view of the present status of parasitic weed problems and their control.

Keywords: parasitic weeds; *Striga*; *Orobanche*; *Phelipanche*

Citation: Parker, C. A Personal History in Parasitic Weeds and Their Control. *Plants* **2021**, *10*, 2249. https://doi.org/10.3390/plants10112249

Academic Editors: Evgenia Dor and Yaakov Goldwasser

Received: 24 September 2021
Accepted: 19 October 2021
Published: 21 October 2021

Publisher's Note: MDPI stays neutral with regard to jurisdictional claims in published maps and institutional affiliations.

1. Introduction

Prior to 1950, the most notable achievement in parasitic weed control had been the discovery in the early 20th century, of immunity to *Orobanche cumana* in sunflower. In their excellent review, Molinero-Ruiz et al. [1] describe how *O. cumana* (then known as *O. cernua*) was first recognised as a problem in sunflowers in the Black Sea region of Russia in 1866, and by the end of the century had spread to Moldova and Romania. Pustovoit [2] describes how, in 1912, a resistance breeding programme was initiated. Cubero [3,4], in other valuable reviews, describes how, by 1916, a range of material had been developed with complete resistance and by 1925, 95% of the sunflower crop grown in Pustovoit's region was based on these resistant lines. In the following three years, however, the resistance failed due to the emergence of different races of the parasite, the original being designated race A and the more virulent, race B. In due course, new varieties were required with resistance to races C, D, E, F, G. and H.

Other parasitic weeds, notably other *Orobanche*, *Phelipanche* and *Striga* problems have proved more intractable. The earliest reports of work on *Striga* as a weed problem are for *S. asiatica* from South Africa where Pearson [5] showed the benefit of nitrogen fertilization and Saunders from 1926 [6] onwards through to 1942, conducted a wide range of studies on biology and control, proposing a number of agronomic approaches, and most importantly began the process of breeding and selection for resistance in sorghum, leading to release of the variety Radar. Timson worked on serious infestations of *S. asiatica* in maize in Rhodesia (Zimbabwe) from 1929 onwards. He promoted catch-cropping with sudan grass, (*Sorghum x drummondii*) together with phosphate fertilizer and crop rotation leading to over 3-fold yield increases in maize over a four-year period [7].

The problem from *S. hermonthica* in the northern half of Africa did not appear to be recognised until somewhat later. Presumably, there was less pressure on land for arable cropping, and it was traditionally suppressed by shifting cultivation. It was first recorded as a problem in maize in Kenya in 1928, and later in sorghum in Sudan, and in rice in Senegal. One of the first to achieve some success in control was Doggett [8], working in Tanzania from 1952 to 1954, who developed a number of resistant sorghum varieties, including Dobbs and others which provided the basis for subsequent work on resistance by the International Crops Research Institute in the Semi-Arid Tropics (ICRISAT) and others.

We have all this early information thanks to the extraordinarily detailed abstracts compiled by McGrath et al., [9] and published by the US Department of Agriculture (USDA) following the discovery of widespread infestation of maize by *S. asiatica* (first reported as *S. lutea*) in the Carolinas of USA, after what must have been a decade unidentified. The fear was that it could spread to the corn belt of the mid-West. This led to the establishment of the Witchweed Lab in Whiteville, North Carolina in 1965, headed by the late Bob Eplee, and a quarantine campaign and series of valuable studies on the biology and control of this species. These included methods for monitoring infestations by separation of seeds from the soil, the use of herbicide to prevent new seed production, and the use of ethylene to stimulate suicidal germination. Much of this work was summarised in the Weed Science Society of America (WSSA) publication by Sand et al. [10]. Wide-scale quarantine restrictions were finally lifted in 2009 and by 2015, $250 million later, it was finally reduced to just over 1000 acres [11].

Beyond the 1950s, major publications across the years have included Job Kuijt's masterly volume on The Biology of Parasitic Flowering Plants, the first to provide an overall survey of the subject [12]. Since then, some of the further major publications have included: Musselman, [13], Visser [14], Press and Graves [15], Joel et al. [16], Heide-Jorgenson [17] Joel et al. [18] and many others have appeared each year.

2. Our studies in the UK

My introduction to parasitic weeds was in the 1950s when I was working for a chemical company in South Africa evaluating their herbicides on field crops, and saw *S. asiatica* for the first time in maize. I did no work on it then but on return to the Agricultural Research Council (ARC) Unit of Experimental Agronomy in Oxford, UK in 1959, I inherited a programme of herbicide evaluation, partly funded by the UK Ministry of Overseas Development (later the Overseas Development Organisation) which already involved work on *S. hermonthica*. From then on, I gradually became more specialised in parasitic weeds, though for most of my career continuing with projects on biology and control of other weeds. There, and subsequently at the ARC Weed Research Organisation (WRO) (Figure 1) to which I was transferred in 1966, we spent many years delving into *Striga* spp., their biology and control, in lab and glasshouse. We screened innumerable herbicides but found none sufficiently selective. In conjunction with the International Crops Research Institute for the Semi-Arid Tropics (ICRISAT), we screened hundreds of sorghum varieties for their stimulant exudation, helping to lead to the standard low-stimulant variety SRN49 [19]. We also found good resistance in semi-wild, 'shibra' millet, but it did not prove practical to exploit this resistance [20].

Figure 1. Weed Research Organisation, Begbroke Hill, Oxford, closed 1985. There were modern labs behind.

We developed a polythene bag technique for testing the effects of nutrients and other studies, allowing ready observation of parasite development [21]. We enjoyed demonstrating the profound inhibitory effect of *Striga* extremely early after host attachment, such that, at 4 weeks, less than 1 mg of *S. hermonthica* seedlings a few mm long could cause 400 mg reduction in the total weight of the host, and at 5 weeks, 13.5 mg caused a total weight loss of 960 mg. Furthermore, the shoot is disproportionately affected as a result of a significant shift in root:shoot balance [22]. We confirmed the effect of nitrogen in reducing infection and of potassium in stimulating it but somehow failed to show the effect of phosphorus in reducing it, now well confirmed by others. It has been shown possible to exploit this beneficial effect of P very economically by 'micro-dosing' [23].

We looked at a stage in parasite infection that has been generally neglected—the role of chemotropism—and found significant evidence for its role in the orientation of *Striga* seedlings towards the host root prior to attachment [24].

A significant date in the history of *Striga* research is when Cook et al. [25] first described the structure of a strigolactone—strigol, from the roots of cotton. There was soon interest in synthesising simpler analogues and Prof. Alan Johnson, director of an ARC research unit at the University of Sussex set a graduate student to the task. He did not immediately have a means of assaying products for their activity and approached Prof Geoffrey Blackman, director of our sister unit at Oxford University (at an ARC dinner) for advice. He said we should be able to help and within a week, we had the first results showing that the very first compounds to be produced had high activity on pre-conditioned *S. hermonthica* seeds [26]. Hence the student Gerald Rosebery achieved instant immortality through his initials, in the form, initially, of GR2, GR3, GR5, GR7, and eventually, in the standard GR24.

Lytton Musselman and I first met in 1973 (at the Malta meeting—see below) and his friendship and support have been pivotal to me ever since. He joined me for an extremely productive sabbatical at WRO in 1980 and we conducted a range of joint studies, on *Striga*, *Phelipanche* and *Orobanche* spp., covering their host ranges, specificity, pollination and seed morphology. Some of the first electron micrographs of the seeds of *Striga* species showed clear specific variations [27]. We also confirmed the narrow host specificity of *S. gesnerioides* strains parasitizing wild species [28]. and the relative lack of host specificity in *O. crenata*, *Phelipanche ramosa* and *P. aegyptiaca*, while *O. minor* showed distinct strains [29]. A pollination study confirmed the strict autogamy of *S. asiatica*, *S. gesnerioides* and *S. forbesii*, also *O. cernua* and *O. minor*, while *S. hermonthica* is dependant on out-crossing. *Orobanche crenata* and *P. aegyptiaca* proved mainly allogamous but facultatively autogamous [30,31].

In 1985 I visited Charlie Riches in Botswana whose work there included the problem of *Alectra vogelii* in cowpea. By then he had identified a number of cowpea landraces with resistance to *Alectra*. I returned to UK with samples of ten of these and in a simple screening to look for possible co-resistance to *S. gesnerioides*, nine out of the ten showed no resistance but B.301 had apparent immunity to both species. In 1984, the variety Suvita-2 had been shown to resist *S. gesnerioides* in Burkina Faso, but it was soon learnt that this line and another, 58–57 were not resistant in Mali, Niger or Nigeria [32]. Our further work with B.301 showed that it was immune to the races from all these countries and from Cameroon [33]. Only in 1993 was it found to be overcome by the 'hyper-virulent' Zakpota race from southern Benin. More recent work is ably reviewed in the paper by Botanga and Timko [34] establishing that there have been other lines identified by the International Institute of Tropical Agriculture (IITA) with broad-spectrum resistance, including to the Zakpota race, but B.301 continued to be valuable to IITA in the development of cowpea lines with dual resistance to both *Striga* and *Alectra*. A study of the host specificity of *A. vogelii* showed there to be at least 3 strains varying in their virulence on cowpea, groundnut and bambara [35]. One further study showed that populations of *A. vogelii* are different in West Africa and southern Africa [36].

One of my final research projects involved a study of the involvement of ethylene in the germination of *Striga*. Ethylene had been used as an important means of stimulating suicidal germination and hence reducing *S. asiatica* seed in the soil in the USA. We wondered to what extent ethylene was involved in the activity of the strigolactones. We confirmed that GR24 greatly stimulated evolution of ethylene in the seeds and significantly increased germination of *S. hermonthica* (the increase was prevented by the ethylene inhibitor, norbornadiene) but was not essential to GR24 activity [37].

3. International Parasitic Plant Meetings

In the early 1970s, Dr Abed Saghir of the American University of Beirut and I established the European Weed Research Council (EWRC) Parasitic Weeds Research Group. We located just over 100 workers on parasitic weeds in 36 countries in Europe and beyond and invited them to attend a Symposium on Parasitic Weeds in Malta in April 1973 (Figure 2). About 50 attended including Lytton Musselman—also Jose Cubero and Job Kuijt but not many others who are still active. The Proceedings were published by EWRC. This meeting arose out of a UK-funded project on *Orobanche crenata* in Malta in faba beans, led by the late Prof Bill Edwards, whom I had visited in 1970.

Figure 2. The Malta participants 1973. (**a**) Chris P; (**b**) Job Kuijt; (**c**) Jose Cubero (**d**) Abed Saghir; (**e**) Lytton M.

After 1973, the EWRC Parasitic Weeds Research Group was taken over by the newly formed European Weed Research Society but it was difficult to give adequate emphasis to *Striga* in a European context, so after a brief divorce, it re-formed in 1979 as the International Parasitic Seed Plant Research Group, which later still was taken under the wing of the International Parasitic Plant Society (IPPS). Lytton Musselman in due course arranged a Second International meeting in Raleigh, North Carolina in 1979. This meeting has been followed by further major meetings around the world every few years—the next, the 16th, to be held in Kenya 2022. The earlier meetings involved the preparation of proceedings ahead of the event, and I was often involved in their editing and publication. I have enjoyed attending most meetings. I sadly missed just two, those in Italy, 2011 and in Amsterdam, 2017, due to ill-health. Figure 3 shows the author with Lytton Musselman in the field after the 10th meeting in Turkey in 2009.

Figure 3. Chris P. and Lytton M. hunting for *Arceuthobium oxycedri* on Mt Sypilos in Turkey 2009.

4. Haustorium

In addition to the major international meetings, there been many more localised or specialised meetings in between, many of them very important and productive, one of particular significance being the *Striga* workshop arranged in Khartoum in 1978 by the International Development Research Center which funded a number of projects on parasitic weeds. It was here that Lytton and I first discussed the idea of a parasitic plants newsletter, resulting in the first issue of 'Haustorium' being published in 1979. It started small and had some lapses and problems of funding documented in the item 'How Haustorium Happens' in our 50th issue [38], but fortunately, it was able to continue and flourish and has endeavoured to briefly note and summarise as much new literature as was readily available on parasitic plants (not just weeds) twice a year. The mailing list for issue 80 currently approaches 500 from some 60 or more countries. We believe that this newsletter has helped foster the widespread interest in parasitic plants, and extensive international collaboration, which culminated in the establishment of the International Parasitic Plant Society.

5. General Publications

In addition to research, my job as 'Tropical Weeds Liaison Officer' involved a substantial amount of survey and advisory work across the world, so I had the opportunity to obtain an overview of the parasitic weed problems in many regions, including the Near East [39]. I was also involved in a World Bank project for three years from 1986 to 1989 in Ethiopia, with the remit to establish a *Striga* research project. In addition to producing research papers of local interest, I had the opportunity to survey other parasitic weed problems across the country [40].

And following my formal retirement in 1990, my colleague Charlie Riches and I were able to consolidate our research and survey experience into the book 'Parasitic Weeds of the World—Biology and Control' published by CAB International [41]. Since then, I have had nothing better to do than bore you with endless review papers—I apologise. Many of them have been requested, including this one, so it is not all my fault.

6. How Do I View the Current Situation?

6.1. Striga Species in Cereals

The current extent of the *Striga* problems in maize and sorghum is far from clear. It may be generalised that in high-input farming, there should not be a serious problem with adequate N and P fertilisation and the advanced varieties of sorghum and maize, which should be available locally. In low-input farming, however, the problems persist and may be getting worse locally. The long search for nitrogen fixation in the cereals could one day be an effective solution but seems to be a bit like nuclear fusion for power generation, forever on the horizon.

Research on the strigolactones has been remarkably productive academically, but sadly their use for application to the soil to stimulate suicidal germination has yet to be proved and registered for practical use.

Significant impact has been achieved at least locally in East Africa, using herbicide-resistant varieties combined with herbicide seed treatment [42] but have not been widely adopted. Likewise, The 'push-pull' technique using *Desmodium* spp. as intercrops (with or without the herbicide seed treatment) [43] has proved valuable in certain farming systems under the right climatic conditions.

Most recently, the 'toothpick' technique, for selective control of *Striga* by a fungal pathogen [44] has been registered for use in East Africa and we look forward to discovering how successful and durable that may prove.

In rice, *Striga* spp. have proved equally intractable, but there has been considerable success with the NERICA varieties developed by the Africa Rice Centre in East Africa. These involve crosses between *Oryza sativa* and the African *O. glaberrima*. These were not developed specifically for *Striga*-resistance but some are proving very effective [45].

In cowpea, *S. gesnerioides*-resistant varieties are available for most areas but, as with *O. cumana* in sunflower, new races continue to occur with extra, or different virulence [46].

Global warming can be expected to result in wider occurrence of *Striga* and other parasitic weed problems [47].

I support the suggestion made elsewhere, for the establishment of a *Striga* (or parasitic weeds) research laboratory—the problems continue to need it.

6.2. Other Genera of Orobanchaceae

Rhamphicarpa fistulosa continues to be of increasing concern in rice, with limited options for control, which include sowing early maturing varieties, and some partially resistant varieties [48].

For *Alectra vogelii*, resistant cowpea varieties are available) [49], while there are no easy solutions in groundnut/peanut

Orobanche cumana in sunflower: (1) reviewed recent progress in the tussle between the breeders and the development of new virulence in sunflower, which I believe the breeders are generally winning. In most regions, resistance is available but the situation keeps changing, keeping the breeders on their toes.

The 2013 workshop in Morocco [50] highlighted the problem from *O. crenata* not only in faba bean, but also in lentil, pea and other legumes across the Mediterranean region, and its serious effects on the nutrition and economy of farmers and countries in the region. In the years since the Malta Symposium there have been many studies on resistance in faba bean. The Egyptian variety Giza 402 was not a productive variety but has been the source of partial resistance in a number of newer varieties. Now a number of varieties are being developed with useful resistance including variety Baraca and derivatives from it, which have shown promise not only for *O. crenata* but also for *O. foetida*, the relatively new problem in Tunisia and Morocco [51].

Meanwhile, *O. crenata* is still spreading. It was first recorded in Ethiopia in 1989, since when it has spread rapidly to many of the important faba-bean growing areas of the country [52].

6.3. Other Parasitic Weed Problems

Other parasitic weed problems that I have become familiar with but have done little or no work on include other broomrapes, *O. cernua* parasitising Solanaceae, and the *Phelipanche* species parasitising Solanaceae, Brassicacae and other families. These present serious continuing problems including *P. ramosa* infestations in rape-seed in France [53]. Control of all these depends mainly on herbicides—some treatments are particularly well developed in Israel.

Cuscuta spp., especially *C. campestris*, can be severe locally on some crops especially when it is a contaminant of crop seed as in lucerne/alfalfa and in niger seed (*Guizotia*

abysssinica), as in Ethiopia. Control depends almost completely on seed-cleaning and on herbicides.

The similar but unrelated *Cassytha filiformis* can be a problem, so far without any developed control measures.

The dwarf mistletoes, *Arceuthobium* species have been described as the most serious disease problem in North American forestry [54]. Their control depends on cultural methods including fire and thinning. Climate change may increase the problem at higher latitudes.

Viscum album causes less severe damage, but there are reports of increasing severity of infestation in conifer crops in Europe, to some extent due to drought [55]. There are no established control methods.

Funding: This research received no external funding.

Institutional Review Board Statement: Not applicable.

Informed Consent Statement: Not applicable.

Data Availability Statement: Not applicable.

Acknowledgments: I wish to acknowledge the help and contributions of my colleagues at Weed Research Organisation; Anne-Marie Hitchcock, Anita Wilson, Della Striger, Nick Dixon, Tess Moore and others. Also the immense help and friendly collaboration that I have enjoyed with Lytton Musselman ever since our first meeting in Malta in 1973. He has had a major role in so many of the activities and achievements described above. He has also assisted me greatly in the writing of this article.

Conflicts of Interest: The author declares no conflict of interest.

References

1. Molinero-Ruiz, L.; Delavault, P.; Pérez-Vich, B.; Pacureanu-Joita, M.; Bulos, M.; Altieri, E.; Domínguez, J. History of the race structure of *Orobanche cumana* and the breeding of sunflower for resistance to this parasitic weed: A review. *Span. J. Agric. Res.* **2015**, *13*, e10R01. [CrossRef]
2. Pustovoit, V.S. *Selection, Seed Culture and Some Agrotechnical Problems of Sunflower*; Translation from Russian, Indian National Scientific Documentation Centre; LH/VPO: Wageningen, The Netherlands, 1976.
3. Cubero, J.I. Breeding for Resistance to Orobanche and Striga: A Review. In Proceedings of the a Workshop on Biology and Control of Orobanche, Wageningen, The Netherlands, 13–17 January 1986; pp. 127–139.
4. Cubero, J.I. Breeding for Resistance to Orobanche Species: A Review. In Proceedings of the International Workshop on Orobanche Research, Obermarchtal, Eberhard-Karls-Universität, Tübingen, Germany, 19–22 August 1989; pp. 257–277.
5. Pearson, H.H.W. *The Problem of the Witchweed*; Publication No. 40; South African Department of Agriculture: Pretoria, South Africa, 1913; 34p.
6. Saunders, A.R. Field experiments at Potchefstroom. A summary of investigations conducted during the period 1903–1940. *S. Afr. Dep. Agric. For. Sci. Bull.* **1942**, *14*, 14–138.
7. Timson, S.D. Witchweed demonstration farm. Final progress report, December 1944. *Rhod. Agric. J.* **1945**, *42*, 404–409.
8. Doggett, H. *1954. Annual Report of the Botanist for the Season 1952–1953*; Tanganyika Department of Agriculture Annual Report; Tanganyika Department of Agriculture, Ukiruguru: Mwanza, Tanganyika, 1953; Part 2; pp. 78–83.
9. McGrath, H.; Shaw, W.C.; Jansen, L.L.; Lipscomb, B.R.; Miller, P.R.; Ennis, W.B. *Witchweed (Striga asiatica)—A new Parasitic Plant in the United States*; USDA Special Publication 10; USDA: Washington, DC, USA, 1957; 142p.
10. Sand, P.F.; Eplee, R.E.; Westbrooks, R.G. (Eds.) *Witchweed Research and Control in the United States of America*; Weed Science Society of America: Champaign, IL, USA, 1990; 154p.
11. Lebrun, A. (Ed.) *Witchweed Eradication Program in North and South Carolina Environmental Assessment, April 2020*; USDA: Washington, DC, USA, 2020. Available online: https://www.aphis.usda.gov/plant_health/ea/downloads/2020/witchweed-north-south-carolina.pdf (accessed on 20 October 2021).
12. Kuijt, J. *The Biology of Parasitic Flowering Plants*; University of California Press: Berkeley, CA, USA, 1969; 248p.
13. Musselman, L.J. The biology of Striga, Orobanche and other root parasitic weeds. *Annu. Rev. Phytopathol.* **1980**, *18*, 483–489. [CrossRef]
14. Visser, J. *South African Parasitic Flowering Plants*; Juta Co.: Cape Town, South Africa, 1981; 184p.
15. Press, M.C.; Graves, J.D. (Eds.) *Parasitic Plants*; Chapman and Hall: London, UK, 1995; 292p.

16. Joel, D.M.; Hershenhorn, Y.; Eizenberg, H.; Aly, R.; Ejeta, G.; Rich, P.J.; Ransom, J.K.; Sauerborn, J.; Rubiales, D. Biology and management of weedy root parasites. In *Horticultural Reviews*; Janick, J., Ed.; Wiley: Hoboken, NJ, USA, 2007; Volume 33, pp. 267–349.

17. Heide-Jorgensen, H. *Parasitic Flowering Plants*; Brill: Leiden, The Netherlands, 2008; 438p.

18. Joel, D.M.; Gressel, J.; Musselman, L.J. (Eds.) *Parasitic Orobanchaceae—Parasitic Mechanisms and Control Strategies*; Springer: Berlin/Heidelberg, Germany, 2013; 513p.

19. Parker, C.; Hitchcock, A.M.; Ramaiah, K.V. The germination of *Striga* species by crop root exudates: Techniques for selecting resistant crop cultivars. In Proceedings of the 6th Asian Pacific Weed Science Society Conference, Jakarta, Indonesia, 11–17 July 1977; Volume 1, pp. 67–74.

20. Parker, C.; Wilson, A.K. *Striga-Resistance Identified in Semi-Wild 'Shibra' Millet (Pennisetum sp.)*; Mededelingen van de Faculteit Landbouwetenschappen, Rijksuniversiteit: Gent, Belgium, 1983; Volume 48, pp. 1111–1117.

21. Parker, C.; Dixon, N.H. The use of polyethylene bags in the culture and study of *Striga* and other organisms on plant roots. *Ann. Appl. Biol.* **1983**, *103*, 485–488. [CrossRef]

22. Parker, C. The influence of *Striga* species on sorghum under varying nitrogen fertilization. In Proceedings of the Third International Symposium on Parasitic Weeds, Aleppo, Syria, 7–10 May 1984; pp. 90–98.

23. Jamil, M.; Charnikhova, T.; Jamil, T.; Ali, Z.; Mohamed, N.E.M.A.; van Mourik, T.; Bouwmeester, H.J. Influence of fertilizer microdosing on strigolactone production and *Striga hermonthica* parasitism in pearl millet. *Int. J. Agric. Biol.* **2014**, *16*, 935–940.

24. Dixon, N.H.; Parker, C. Aspects of the resistance of sorghum varieties to *Striga* species. In Proceedings of the Third International Symposium on Parasitic Weeds, Aleppo, Syria, 7–10 May 1984; pp. 123–132.

25. Cook, C.E.; Whichard, L.P.; Wall, M.E.; Egley, G.H.; Coggan, P.; Luhan, P.A.; McPhail, A.T. Germination stimulants 2. The structure of strigol—A potent seed germination stimulant for witchweed (*Striga lutea* Lour.). *J. Am. Chem. Soc.* **1972**, *94*, 6198–6199. [CrossRef]

26. Johnson, A.W.; Rosebery, G.; Parker, C. A novel approach to *Striga* and *Orobanche* control using synthetic germination stimulants. *Weed Res.* **1976**, *16*, 223–227. [CrossRef]

27. Musselman, L.J.; Parker, C. Surface features of *Striga* seeds (Scrophulariaceae). *Adansonia* **1981**, *20*, 431–438.

28. Musselman, L.J.; Parker, C. Preliminary host ranges of some strains of economically important broomrapes (*Orobanche*). *Econ. Bot.* **1982**, *36*, 270–273. [CrossRef]

29. Musselman, L.J.; Parker, C. Studies on indigo witchweed, the American strain of *Striga gesnerioides* (Scrophulariaceae). *Weed Sci.* **1981**, *29*, 594–596. [CrossRef]

30. Musselman, L.J.; Parker, C.; Dixon, N. Notes on autogamy and flower structure in agronomically important species of *Striga* (Scrophulariaceae) and *Orobanche* (Orobanchaceae). *Beiträge Biol. Pflanz.* **1982**, *56*, 329–343.

31. Musselman, L.J.; Parker, C. Biosystematic research in the genus Striga (Scrophulariaceae). In Proceedings of the Second International Striga Workshop, Ougadougou, Burkina Faso, 5–8 October 1981; pp. 19–24.

32. Aggarwal, V.D.; Haley, S.D.; Brockman, F.E. Present status of breeding cowpea for resistance to *Striga* at IITA. In Proceedings of the Workshop on Biology and Control of Orobanche, Wageningen, The Netherlands, 13–17 January 1986; pp. 176–180.

33. Parker, C.; Polniaszek, T.I. Parasitism of cowpea by *Striga gesnerioides*: Variation in virulence and discovery of a new source of host resistance. *Ann. Appl. Biol.* **1990**, *116*, 305–311. [CrossRef]

34. Botanga, C.J.; Timko, M.P. Phenetic relationships among different races of *Striga gesnerioides* (Willd.) Vatke from West Africa. *Genome* **2006**, *49*, 1351–1365. [CrossRef]

35. Polniaszek, T.I.; Parker, C. Variation in Host Specificity of *Alectra vogelii* (Benth. Scrophulariaceae). In Proceedings of the Fourth International Symposium on Flowering Plants, Marburg, Germany, 2–7 August 1987; pp. 613–619.

36. Polniaszek, T.I.; Parker, C.; Riches, C.R. Variation in the virulence of *Alectra vogelii* populations on cowpea. *Trop. Pest Manag.* **1991**, *37*, 152–154. [CrossRef]

37. Jackson, M.B.; Parker, C. Induction of germination by a strigol analogue requires ethylene action in *Striga hermonthica* but not in *S. forbesii*. *J. Plant Physiol.* **1991**, *138*, 383–386. [CrossRef]

38. Parker, C.; Musselman, L.J. How Haustorium happens. *Haustorium* **2006**, *50*, 2–4.

39. Parker, C.; Wilson, A.K. Parasitic weeds and their control in the Near East region. *FAO Plant Prot. Bull.* **1986**, *34*, 83–93.

40. Parker, C. Parasitic plants in Ethiopia. *Walia* **1988**, *11*, 21–27.

41. Parker, C.; Riches, C.R. *Parasitic Weeds of the World: Biology and Control*; CAB International: Wallingford, UK, 1993; 340p.

42. Kabambe, V.H.; Kanampiu, F.; Ngwira, A. Imazapyr (herbicide) seed dressing increases yield, suppresses *Striga asiatica* and has seed depletion role in maize (*Zea mays* L.) in Malawi. *Afr. J. Biotechnol.* **2008**, *7*, 3293–3298.

43. Kanampiu, F.; Makumbi, D.; Mageto, E.; Omanya, G.; Waruingi, S.; Musyoka, P.; Ransom, J. Assessment of management options on *Striga* infestation and maize grain yield in Kenya. *Weed Sci.* **2018**, *66*, 516–524. [CrossRef] [PubMed]

44. Sands, D.; Baker, C. Parasites of parasites—The toothpick project. *Haustorium* **2020**, *78*, 5–6.

45. Rodenburg, J.; Cissoko, M.; Kayeke, J.; Dieng, I.; Khan, Z.R.; Midega, C.A.O.; Onyuka, E.A.; Scholes, J.D. Do NERICA rice cultivars express resistance to *Striga hermonthica* (Del.) Benth. and *Striga asiatica* (L.) Kuntze under field conditions? *Field Crop. Res.* **2015**, *170*, 83–94. [CrossRef] [PubMed]

46. Ohlson, E.W.; Timko, M.P. Race structure of cowpea witchweed (*Striga gesnerioides*) in West Africa and its implications for *Striga* resistance breeding of cowpea. *Weed Sci.* **2020**, *68*, 125–133. [CrossRef]

47. Mohamed, K.I.; Papes, M.; Williams, R.; Benz, B.W.; Peterson, A.T. Global invasive potential of 10 parasitic witchweeds and related Orobanchaceae. *Ambio* **2006**, *35*, 281–288. [CrossRef] [PubMed]

48. Rodenburg, J.; Cissoko, M.; Dieng, I.; Kayeke, J.; Bastiaans, L. Rice yields under *Rhamphicarpa fistulosa*-infested field conditions, and variety selection criteria for resistance and tolerance. *Field Crop. Res.* **2016**, *194*, 21–30. [CrossRef]

49. Dieni, Z.; Tignegre, J.B.; De La SalleTignerge, J.-B.; Tongoona, P.; Dzidzienyo, D.; Asante, I.K.; Ofori, K. Identification of sources of resistance to *Alectra vogelii* in cowpea (*Vigna unguiculata* (L.) Walp.) germplasm from Burkina Faso. *Euphytica* **2018**, *214*, 234. [CrossRef]

50. Parker, C.; O'Sullivan, D. Building a new research alliance to reclaim faba bean production area abandoned to *Orobanche*. *Haustorium* **2013**, *64*, 4–6.

51. Rubiales, D.; Flores, F.; Emeran, A.A.; Kharrat, M.; Amri, M.; Rojas-Molina, M.M.; Sillero, J.C. Identification and multi-environment validation of resistance against broomrapes (*Orobanche crenata* and *Orobanche foetida*) in faba bean (*Vicia faba*). *Field Crop. Res.* **2014**, *166*, 58–65. [CrossRef]

52. Teklay, A.; Hadas, B.; Yemane, N. Distribution and economic importance ofbroomrape (*Orobanche crenata*) in food legumes production of south Tigray, Ethiopia. *J. Crop Prod.* **2013**, *2*, 101–106.

53. Gibot-Leclerc, S.; Pinochet, X.; Salle, G. Orobanche rameuse (*Orobanche ramosa* L.) du colza: Un risque émergent sous surveillance (Broomrapes (*Orobanche ramosa*) as parasite of winter oilseed rape: An extension risk under observation). *OCL Ol. Corps Gras Lipides* **2006**, *13*, 200–205. [CrossRef]

54. Hawksworth, F.G.; Wiens, D. *Dwarf Mistletoes: Biology, Pathology, and Systematics*; Agriculture Handbook 709; United States Department of Agriculture Forest Service: Washington, DC, USA, 1996; 410p.

55. Férriz, M.; Martin-Benito, D.; Cañellas, I.; Gea-Izquierdo, G. Sensitivity to water stress drives differential decline and mortality dynamics of three co-occurring conifers with different drought tolerance. *For. Ecol. Manag.* **2021**, *486*, 118964. [CrossRef]

plants

MDPI

Article

Characterization of a Chickpea Mutant Resistant to *Phelipanche aegyptiaca* Pers. and *Orobanche crenata* Forsk

Shmuel Galili [1,*], Joseph Hershenhorn [2], Evgeny Smirnov [2], Koichi Yoneyama [3], Xiaonan Xie [3], Orit Amir-Segev [1], Aharon Bellalou [1] and Evgenia Dor [2,*]

1 Institute of Plant Sciences, The Volcani Center, Agricultural Research Organization, P.O. Box 15159, HaMaccabim Road 68, Rishon LeZion 7505101, Israel; oritas@volcani.agri.gov.il (O.A.-S.); aharonb@volcani.agri.gov.il (A.B.)
2 Institute of Plant Protection, Newe Ya'ar Research Center, Agricultural Research Organization, P.O. Box 1021, Ramat Yishay 3009503, Israel; hershenj@gmail.com (J.H.); evgeny@volcani.agri.gov.il (E.S.)
3 Center for Bioscience Research and Education, Utsunomiya University, 350 Mine-machi, Utsunomiya 321-8505, Japan; yone2000@sirius.ocn.ne.jp (K.Y.); xie@cc.utsunomiya-u.ac.jp (X.X.)
* Correspondence: galilis@agri.gov.il (S.G.); evgeniad@volcani.agri.gov.il (E.D.)

Citation: Galili, S.; Hershenhorn, J.; Smirnov, E.; Yoneyama, K.; Xie, X.; Amir-Segev, O.; Bellalou, A.; Dor, E. Characterization of a Chickpea Mutant Resistant to *Phelipanche aegyptiaca* Pers. and *Orobanche crenata* Forsk. *Plants* **2021**, *10*, 2552. https://doi.org/10.3390/plants10122552

Academic Editor: Alexandra S. Dubrovina

Received: 15 August 2021
Accepted: 17 November 2021
Published: 23 November 2021

Publisher's Note: MDPI stays neutral with regard to jurisdictional claims in published maps and institutional affiliations.

Abstract: Chickpea (*Cicer arietinum* L.) is a major pulse crop in Israel grown on about 3000 ha spread, from the Upper Galilee in the north to the North-Negev desert in the south. In the last few years, there has been a gradual increase in broomrape infestation in chickpea fields in all regions of Israel. Resistant chickpea cultivars would be simple and effective solution to control broomrape. Thus, to develop resistant cultivars we screened an ethyl methanesulfonate (EMS) mutant population of F01 variety (Kabuli type) for broomrape resistance. One of the mutant lines (CCD7M14) was found to be highly resistant to both *Phelipanche aegyptiaca* and *Orobanche crenata*. The resistance mechanism is based on the inability of the mutant to produce strigolactones (SLs)—stimulants of broomrape seed germination. LC/MS/MS analysis revealed the SLs orobanchol, orobanchyl acetate, and didehydroorobanchol in root exudates of the wild type, but no SLs could be detected in the root exudates of CCD7M14. Sequence analyses revealed a point mutation (G-to-A transition at nucleotide position 210) in the Carotenoid Cleavage Dioxygenase 7 (CCD7) gene that is responsible for the production of key enzymes in the biosynthesis of SLs. This nonsense mutation resulted in a CCD7 stop codon at position 70 of the protein. The influences of the CCD7M14 mutation on chickpea phenotype and chlorophyll, carotenoid, and anthocyanin content were characterized.

Keywords: chickpea; strigolactone; broomrape resistance; chickpea phenotype; chlorophyll; carotenoid; anthocyanin

1. Introduction

Chickpea (*Cicer arietinum* L.) is an important legume crop grown on over 10 million ha in at least 37 countries worldwide, including India (65%), Pakistan (10%), Iran (8%), and Turkey (5.5%). [1]. In Israel chickpea is one of the main legume crops, grown on about 3000 ha with an average yield of about 3.5 t/ha. In recent years, the two broomrape species in Israel, Egyptian broomrape (*Phelipanche aegyptiaca* Pers.) and crenate broomrape (*Orobanche crenata* Forsk.), have become a major problem in chickpea field production [2]. The only broomrape-control methods that have been successfully utilized commercially in other crops are resistant varieties and chemical control [3–5].

Broomrapes (*Phelipanche* spp. and *Orobanche* spp.) are worldwide weedy root parasites of dicotyledonous crops, causing severe losses in the yield and quality of agricultural crops [6,7]. The initial step of broomrape–plant recognition involves root-exuded strigolactones (SLs), which have long been known to induce broomrape seed germination [8,9], and have been recently recognized as plant hormones affecting plant development and growth [10]. SLs consist of a tricyclic lactone (A, B, and C rings) connected to a butenolide group (D ring) via an enol ether bridge. SLs' degree of activity, function, and specificity depend on the various substituents on the A and B rings [11]. The SL biosynthesis pathway in plants is derived from the carotenoid pathway [12–14], in which β-carotene is converted into carlactone by three catalytic enzymes: D-27 (9-cis/all-*trans*-β-carotene isomerase) [15], and two carotenoid cleavage dioxygenases, CCD7 and CCD8 [16,17]. Carlactone is converted to SLs by the cytochrome P450 monooxygenase- homolog activity of MORE AXILLARY GROWTH1 (MAX1) in rice [18], and MAX1 and lateral branching oxidoreductase in *Arabidopsis* [16,19,20]. SLs are produced mainly in roots and their active transport to the rhizosphere by the exporter pleiotropic drug resistance 1 (PDR1), identified in *Petunia*, was shown [21–23].

In a previous study, we obtained a tomato CCD7-deletion mutant showing broomrape resistance [24,25]. SL-deficient sorghum and rice mutants also demonstrate high degrees of resistance to *Striga* spp. [26,27]. Moreover, resistance to parasitic weeds based on low SL exudation exists in pea and faba bean germplasms [28,29]. Mutants defective in SL biosynthesis are characterized by a highly branched/tillering phenotype [30,31]. Furthermore, SLs regulate root architecture [16,32–35].

The objectives of the present study were to isolate and characterize an ethyl methanesulfonate (EMS)-mutagenized F01 chickpea mutant, CCD7M14, which shows considerable resistance to broomrape, and to elucidate its resistance mechanism, characterize its phenotype, and determine its leaf chlorophyll, carotenoid and anthocyanin contents.

2. Results

2.1. Mutagenesis and Screening for Broomrape Resistance

EMS mutagenesis was applied to seeds of a wild-type (WT) F01 chickpea breeding line (Kabuli type), and 3000 families of the second generation were tested for resistance to both *P. aegyptiaca* and *O. crenata*. A chickpea mutant showing high resistance to both broomrapes, was identified—CCD7M14 (Figure S1).

2.2. Phenotyping

2.2.1. Resistance to *P. aegyptiaca* and *O. crenata*

P. aegyptiaca shoots began to emerge aboveground 8 weeks after sowing in pots with WT F01 plants. At this time, about 20% of the WT F01 plants were infected with one or two shoots (Figure 1a). Both shoot number above the soil and percentage of infected plants increased rapidly over time, and at the end of the experiment (14 weeks), all WT F01 plants were infected with 8–10 aboveground shoots. At this time only one broomrape shoot was observed in two pots planted with CCD7M14 (percentage of infected plants was 20%). Throughout the course of the experiment, both percentage of infected plants and number of aboveground shoots per plant were significantly lower for the mutant plants. The roots were washed and broomrape number and biomass were recorded. About 16.10 ± 4.23 broomrape shoots were counted per WT F01 plant with average biomass of 82.11 ± 6.69 g, whereas only 1.60 ± 1.78 shoots with total biomass of 7.93 ± 5.15 g were found per mutant plant (Table 1).

O. crenata developed more slowly than *P. aegyptiaca*. First *O. crenata* shoots emerged aboveground 12 weeks after planting in WT F01 pots (Figure 1b). At the end of the experiment (20 weeks after sowing), 90% of WT F01 plants were infected with one or two shoots. About 13.6 ± 3.48 broomrapes with a total biomass of about 109.74 ± 10.92 g per WT F01 plant were observed after root washing (Table 1). CCD7M14 plants were highly resistant to *O. crenata*. Only one aboveground shoot was observed in one pot at the end of the experiment, and about 2.20 ± 1.71 broomrapes with a total biomass of 13.78 ± 6.96 g were counted on the washed roots (Table 1).

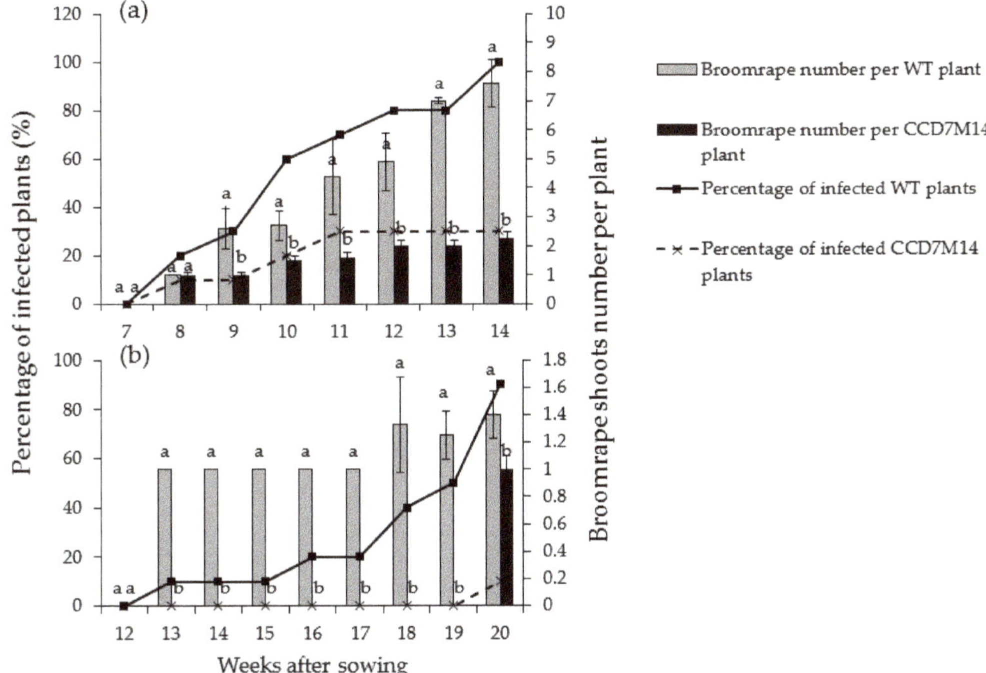

Figure 1. Aboveground broomrape shoots in pots planted with WT F01 or the CCD7M14 mutant. The experiments were arranged in a completely randomized design with 10 replications (pots) per treatment. Lines show the percentages of infected plants, bars indicate the average numbers of aboveground shoots attached to the infected plants. (**a**) Infestation with *P. aegyptiaca*. (**b**) Infestation with *O. crenata*. Vertical lines indicate standard error of the mean (SEM). Lowercase letters indicate least-significant differences (LSD), based on the Tukey–Kramer honestly significant difference test ($\alpha = 0.05$) between the chickpea lines.

Table 1. Statistical analysis of the chickpea resistance experiments. The results were subjected to ANOVA. The experiments were conducted with ten replications. SEM—standard error of the mean, dF—Degrees of Freedom; F—F ratio, Prob > F— F probability.

Parameter	Chickpea Line	Average Mean	SEM	dF	F	Prob > F
		P. aegyptiaca				
Broomrape number	WT F01	16.1	4.23	1	99.96	<0.0001
	CCD7M14	1.6	1.78			
Broomrape biomass (g)	WT F01	82.11	6.69	1	77.17	<0.0001
	CCD7M14	7.93	5.15			
		O. crenata				
Broomrape number	WT F01	13.6	3.48	1	9.6	0.0062
	CCD7M14	2.2	1.71			
Broomrape biomass (g)	WT F01	109.74	10.92	1	54.92	<0.0001
	CCD7M14	13.78	6.96			

2.2.2. Resistance Mechanism

To determine whether the resistance mechanism of CCD7M14 was based on its inability to synthesize SLs or secrete them into the rhizosphere, we tested its ability to stimulate broomrape seed germination. *P. aegyptiaca* seed germination near the WT F01 root system was high (76.84 ± 6.28%), whereas in the pots with CCD7M14 plants, only 0.72 ± 0.45% of the seeds germinated. Germination of *O. crenata* seeds was about 42.12 ± 2.57% in the pots with the WT, whereas in the mutant pots, no *O. crenata* seed germination was observed (Table 2).

Table 2. Statistical analysis of the broomrape seed germination closed to chickpea roots. The results were subjected to ANOVA. The experiments were conducted with five replications. SEM—standard error of the mean, dF—Degrees of Freedom; F—F ratio, Prob > F—F probability.

Broomrape	Chickpea Line	Average Mean	SEM	dF	F	Prob > F
P. aegyptiaca	WT F01	76.84	6.28	1	146.00	<0.0001
	CCD7M14	0.72	0.45			
O. crenata	WT F01	42.21	2.57	1	270.07	<0.0001
	CCD7M14	0	0			

The ability of WT and CCDM14 root exudates to stimulate *P. aegyptiaca* seed germination was tested in vitro in Petri dishes. Root exudate of the WT applied to the seeds at concentrations of 0.1, 1 and 10 µL/mL caused *P. aegyptiaca* germination at rates of 28.1 ± 5.78, 77.38 ± 3.13, and 84.84 ± 4.28%, respectively (compared to 79.19 ± 1.7% following application of 10^{-6} M GR24, a synthetic SL, as a positive control) (Table 3). A low percentage of seed germination was induced by the mutant root exudates (9.02 ± 0.77, 15.94 ± 1.19, and 34.95 ± 2.52% at concentrations of 0.1, 1 and 10 µL/mL, respectively), but only short radicals developed, which did not continue to elongate normally and were dead after 1 week.

Analysis of SLs in root exudates of WT F01 and CCD7M14 plants revealed the presence of orobanchol, orobanchyl acetate, and putative didehydroorobanchol in WT F01 root exudates, but no SLs in the mutant root exudates (Figure S2).

Table 3. Statistical analysis of the broomrape seed germination caused by root exudates. The results were subjected to ANOVA. The experiments were conducted with five replications. SEM—standard error of the mean, dF—Degrees of Freedom; F—F ratio, Prob > F—F probability.

Root Exudates Concentration (µL/mL)	Chickpea Line	Average Mean	SEM	dF	F	Prob > F
1	WT F01	28.10	5.78	1	10.71	0.0307
	CCD7M14	9.02	0.77			
10	WT F01	77.38	3.13	1	336.71	<0.0001
	CCD7M14	15.94	1.19			
100	WT F01	84.84	4.28	1	100.95	0.0006
	CCD7M14	34.95	2.52			

2.2.3. Plant Morphology and Pigment Contents

CCD7M14 plants had a SL-deficiency phenotype, with a high number of short branches compared to WT F01 plants. No significant differences in foliage or root biomass were found between the lines (Table 4). The CCD7M14 plants had 83% more primary branches than the WT F01 plants, and the mutant's primary branch length was only 66% of that of the WT F01 plant. These morphological changes in CCD7M14 were observed both in the net house and under field conditions (Figure 2a–d), leading to a bushy shape at plant maturity.

Table 4. Morphological characteristics of CCD7M14 compared to WT F01 chickpea.

Parameters	Chickpea Line	
	WT F01	**CCD7M14**
Foliage biomass (g)	242.8 ± 14.5 a	210.2 ± 12.5 a
Root biomass (g)	112.3 ± 8.3 a	113.8 ± 37.2 a
Primary branch number	7.0 ± 0.8 b	12.0 ± 1.4 a
Primary branch length (cm)	62.6 ± 2.0 a	40.3 ± 4.0 b

Data are presented as average mean of 5 replications with standard error of the mean (SEM). Lowercase letters indicate significant differences between the WT F01 and CCD7M14 according to LS means contrast test (α = 0.05).

Figure 2. Morphological differences between WT F01 and CCD7M14. (**a**) One-month-old WT F01 (right) and CCD7M14 (left) plants grown in a net house. (**b1,b2**) Stem distribution on the lower section of the plants. (**c**) Primary branches of WT F01 (left) and CCD7M14 (right) plants. (**d**) Three-month-old WT F01 (left) and CCD7M14 (right) plants in the field.

Analysis of carotenoid, chlorophyll, and anthocyanin contents in the first, third, and fifth leaves revealed significant decreases in carotenoids and chlorophylls, and an increase in anthocyanins in the mutant as compared to its parental line (Table 5).

Table 5. Contents of carotenoids, chlorophyll, and anthocyanins (µg per 1 g of fresh leaf biomass) in the leaves of WT F01 and CCD7M14.

Pigment	Leaf 1		Leaf 3		Leaf 5	
	WT F01	**CCD7M14**	**WT F01**	**CCD7M14**	**WT F01**	**CCD7M14**
Chlorophyll a	214.5 ± 9.2 a	120.3 ± 15.1 b	230.0 ± 30.0 a	157.2 ± 9.5 b	281.5 ± 48.9 a	151.1 ± 30.6 b
Chlorophyll b	183.9 ± 7.4 a	56.0 ± 11.7 b	184.3 ± 47.9 a	74.5 ± 5.0 b	200.0 ± 35.3 a	65.2 ± 17.2 b
Total chlorophyll	402.2 ± 11.5 a	176.3 ± 26.5 b	414.43 ± 80.5 a	231.7 ± 6.1 b	481.6 ± 56.7 a	216.3 ± 53.2 b
Carotenoids	62.1 ± 5.3 a	35.1 ± 4.0 b	66.5 ± 12.1 a	35.8 ± 2.8 b	61.3 ± 8.1 a	27.2 ± 4.5 b
Anthocyanin	9.7 ± 0.9 b	33.2 ± 5.1 a	15.2 ± 2.1 b	32.6 ± 7.3 a	11.9 ± 1.2 b	29.3 ± 4.2 a

Results are presented as average mean of 3 replications with standard error of the mean (SEM). Lowercase letters indicate significant differences between the WT F01 and CCD7M14 according to LS means contrast test (α = 0.05).

2.3. DNA Analysis

Blast analyses of the chickpea genome based on the tomato *CCD7* sequence revealed a single *CCD7* gene with 64.9% protein sequence identity to the tomato protein (Figure 3). DNA sequence analysis of the *CCD7* gene in CCD7M14 compared to the WT F01 line revealed a single G-to-A nucleotide transition at position 210 (Figure 4). This mutation led to stop-codon formation (*) instead of tryptophan (W) at amino acid position 70 (84 in tomato) (Figure 5). No other mutations were found in the chickpea *CCD7* gene.

```
1     MQAKPIHNTPPYIIPPLIRAS.PVHQASLPQTLTKPKPITMPIPNTDVLVPDRPVPVRPF
      ||||:.||.  . .|.|.|  |   | | ||.:..:..:|..|.       |..
18    MQAKACHNINNIPPKLLPPAKLPSTVAMSPSQLTLPSHVARAITITTS.....PTHEVYT

                     *
60    IEHDDSIDASWDYNFLFMSQRSETNQPITLRVVEGAIPIDFPSGAYYLTGPGIFMDDHGS
      | ||.:.| |||.|||:|||||..:|:.|||||||.|| |||||.||||||||:| |||||
73    PEIDDTVTAYWDYQFLFVSQRSEATEPVSLRVVEGSIPSDFPSGTYYLTGPGLFADDHGS

120   TVHPLDGHGYLRAFTFDNVNKKVKYMAKYVQTEAQLEEYDRKTDSWKFTHRGPFSLLKGG
      ||||||||||||.|.:|. . .||:||:|::||||  || |. .:.|:||||||||:||||
133   TVHPLDGHGYLRTFEIDGSTGQVKFMARYIETEAQTEERDPVSGKWRFTHRGPFSVLKGG

180   KRIGNTKVMKNVANTSVLMWGKKLLCMWEGGNPYEIESGTLDTIGKFNMMDGCDMEDHEE
      |.:|||||||||||||| || :|:|:|||||:||||:| ||:|:|||::..| |  .:.
193   KMVGNTKVMKNVANTSVLQWGGRLFCLWEGGDPYEIDSKTLNTLGKFELIKNSDQVLEDK

240   NHGG.DVWEVAANLIKPILYGIFKMPPKRLLSHYKVDSIRNRLLTVSCNAEDMLLPRSNF
      . :  |.::|||.|:|||||||:|||.||||||||:|.  ||||||.:||||||||||||||
253   KISHSDFLDVAAQLLKPILYGVFKMSPKRLLSHYKIDTRRNRLLIMSCNAEDMLLPRSNF

299   TFNEYDSNFKLVEKQEFRIPDHLMIHDWAFTDTHYILFANRIKLDVLGSMAAVCGASPMI
      || |:||||.|::.|||  ||||||||||||||||||:|||||: |||.|||| ||||
313   TFYEFDSNFQLLQSQEFEIPDHLMIHDWAFTDTHYILFGNRIKLDIPGSMTAVCGLSPMI

359   ASLTVNPSKSTSPIYLLPRFPNKLEGQKRDWRIPVEAPSQLWLLHVGNAFEVKQSH.GNL
      ..|.||||||.||||||||.|.  :  |||| |:||||||:|:|||||||||  :.: |||
373   SALSVNPSKPTSPIYLLPRFRNNNVE..RDWRKPIEAPSQMWVLHVGNAFEEIDEQNGNL

418   DIQIQATACSYQWFNFHKLFGYNWQTRQLDPSIMNVKGGNE.LLPHLVQVSIKLDSDYNC
      :||||||.:||||||||:|:|||:||. .||||:|||.:|:| |||||||||.|.||.. ||
431   NIQIQASGCSYQWFNFQKMFGYDWQSGKLDPSMMNVEEGEEKLLPHLVQVCINLDKKGNC

477   QECDVKPIQ.EWKKSSDFPATNPTFSGKKNKYIYAATTLGSRKSLPSFPFDSIVKLDLVN
      .|.|..:. ||.|..|||| ||.|||:||:||||||. |||..|| ||||.:|||: |:
491   TKCSVNDLNPEWNKAADFPAMNPEFSGRKNRYIYAATCTGSRQALPHFPFDAVVKLNAVD

536   NSVHTWTAGSRRFVGEPIFVPKGQE.EDDGYILVVEYAVSMQRCCLVILNPKMIGTKNAL
      .||:.|.||.|||:|||:|:|:|:| : |||||:||||||||| |||.||||:::..||.||.:
551   KSVQKWSAGRRRFIGEPVFIPRGTNKEDDGYLLVVEYAVSTQRCYLVILDAQKIGEKNEV

595   VARLEIPMNLNFPLGFHGFWAASES
      |||||:|.:||||||||||||:.:|
611   VARLEVPRHLNFPLGFHGFWAPTNS
```

Figure 3. Chickpea and tomato CCD7 protein sequence homology. The upper and lower sequences are of chickpea and tomato CCD7, respectively. Identical amino acids are indicated by a solid line, and similar amino acids are indicated by one or two dots according to their similarity levels. The mutated amino acid in CCD7 of broomrape-resistant line CCD7M14, at position 70 (84 according to tomato) is indicated in yellow and marked by an asterisk.

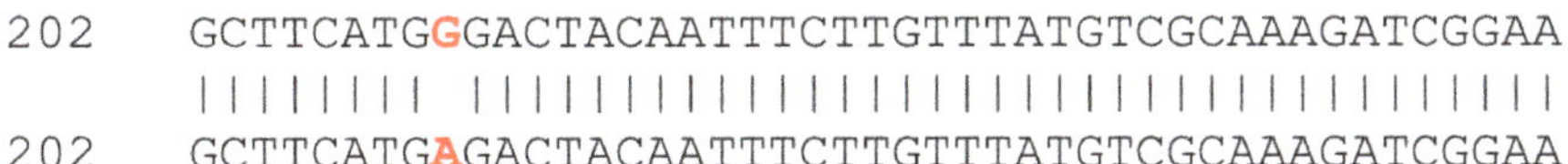

Figure 4. The Blast results of DNA sequences (nucleotides 202–246) of WT F01 chickpea (upper line) and CCD7M14 (lower line). The G-to-A transition at position 210 is indicated in bold red letters.

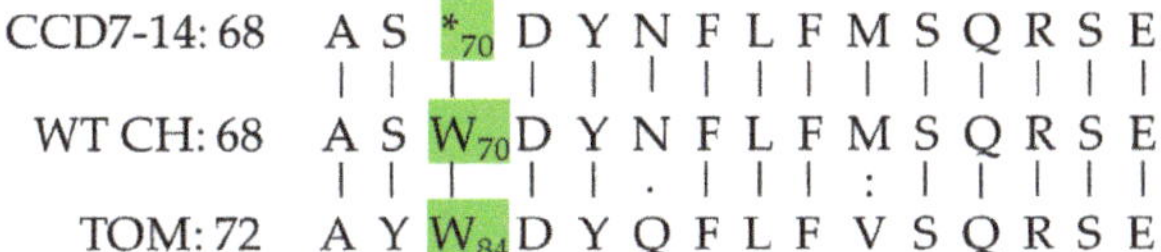

Figure 5. The Blast results of the protein sequences (amino acids 68–82 (82–96 in tomato)) of CCD7M14 (CCD7-14), F01 (WT CH), and tomato (TOM). The W-to-stop codon (*) transition at position 70 (84 in tomato) is indicated in green.

3. Discussion

Chickpea mutant CCD7M14 was produced by EMS mutagenesis. The mutant showed high resistance to both *P. aegyptiaca* and *O. crenata* (Figure S1). Only one mutant plant was infected with a single *P. aegyptiaca*, and one with a single *O. crenata* shoots in all experiments, compared to 90–100% infection in WT F01 plants with more than 8–10 aboveground broomrape shoots (Figure 1a,b). However, once an attachment formed on the mutant roots, parasite development progressed normally. Since no *P. aegyptiaca* or *O. crenata* seed germination was observed near CCD7M14 roots, and its root exudates did not stimulate their seed germination in Petri dishes, it is suggested that the CCD7M14 resistance mechanism is based on its inability to synthesize SLs or to secrete them into the rhizosphere. Indeed, DNA sequence analysis of the CCD7M14 *CCD7* gene revealed stop-codon formation due to a single G-to-A nucleotide transition at position 210 (Figures 4 and 5). This resulted in the absence of the SLs orobanchol, orobanchyl acetate, and didehydroorobanchol in the root exudates (Figure S2), rendering the mutant plant resistant to the parasite because no seed germination could occur near its roots. This resistance mechanism has been reported in tomato [25,36–38], pea [39] and faba bean [30,40]. Previously, this type of resistance had been obtained by fast-neutron mutagenesis [24,25] and targeted mutagenesis [37,38]. It had also been found in wild tomato species (*Solanum pennellii* [36]), and recognized in resistant cultivars of faba bean and pea [30,39,40]. In our case, the resistance was obtained by EMS mutagenesis, where one point mutation in the *CCD7* gene resulted in the formation of a stop codon, leading to the same results as *CCD7* deletion by fast-neutron mutagenesis [25,26] or silencing of *CCD8* using CRISPR/Cas9-mediated mutagenesis [37,38]. It is important to note that to date, all identified *CCD7* genes have been single copies, in contrast to two, four and six copies of *CCD8* identified in maize, rice and sorghum, respectively [41].

It has been shown that plants exude mixtures of several SLs, and every plant species is characterized by a specific SL profile [42]. In the current study, we first identified the SLs produced by chickpea roots. LC/MS/MS analysis revealed that the WT F01 chickpea cultivar produces three SLs: orobanchol, orobanchyl acetate, and putative didehydroorobanchol isomer(s). All three belong to the orobanchol type, which only differs from the strigol-type SLs in the stereochemistry of the C-ring [43], and are derived from 4-deoxyorobanchol in rice [18]. Some other species, such as *Populus*, pea, petunia, and tomato, have been reported to have only orobanchol-type SLs [44]. Orobanchol, first isolated from red clover (*Trifolium pratense* L.) root exudates [9], is probably the most abundant hydroxy-SL in the plant kingdom [42]. This SL assumes to be a central intermediate in SL biosynthesis, and it has been suggested as a precursor of other SL molecules, such as: fabacol, orobanchyl acetate, solanacol, and so on [43]. Putative didehydroorobanchol has been detected in root exudates of tomato [26], tobacco [42], and *Medicago truncatula* [45]; and orobanchyl acetate in red clover [46], rice and tobacco [47]. The didehydroorobanchol isomer in *M. truncatula* was named medicaol [45]. Both orobanchol and orobanchyl acetate have been reported to be produced by Asteraceae plants and by faba bean [48,49].

CCD7M14 was characterized by a typical SL-deficient phenotype—increased branching and reduced plant height. These results are in agreement with Vogel et al. [50], where transgenic tomato plants expressing their endogenous *CCD7* gene in the antisense form also displayed increased branching and reduced plant height. Similar observations have also been reported for pea [51], petunia [52], poplar [53], and *Arabidopsis* [54]. According to Boyer et al. [11], orobanchyl acetate and 5- deoxystrigol are more active at inhibiting shoot branching than strigol and orobanchol. Furthermore, blockage of orobanchol biosynthesis from carlactonoic acid in tomato did not rescue the branching phenotype [55]. The absence of orobanchyl acetate in CCD7M14 plants likely explains its bushy shape at maturity (Figure 5). Orobanchol and putative didehydroorobanchol may be involved in other biological processes, such as regulation of photosynthesis and pigment accumulation. We found significant decreases in chlorophyll and carotenoid contents and an increase in anthocyanins in the leaves of CCD7M14 as compared to the WT F01 line (Table 2). Exogenous application of the synthetic SL GR24 under stress conditions has been shown to control chlorophyll degradation and maintain the photosynthetic rate [56–59]. On the other hand, chlorophyll content in sunflower leaves was not influenced by GR24 treatment of achene pre-sowing, but carotenoid content increased [60]. GR24 has been found to affect ABA-induced activation of anthocyanin biosynthesis in grapevine berries [61]. In transgenic tobacco lines impaired in SL biosynthesis, overaccumulation of anthocyanins in the mature stems likely results from antagonism between the SL and jasmonic acid pathways [62]. SL regulation of anthocyanin accumulation has been shown in *Arabidopsis* [63,64].

4. Materials and Methods

4.1. Plant Material

All experiments were carried out with: (a) a WT F01 chickpea breeding line (Kabuli type) that is erect, produces high yields, and is resistant to both *Fusarium* wilt and *Ascochyta* blight and (b) CCD7M14, a chickpea EMS mutant line derived from WT F01. Broomrape seeds were collected from *P. aegyptiaca* and *O. crenata* inflorescences parasitizing tomato plants grown in Kibbutz Bet Ha'shita (32°33′15″ N 35°26′15″ E) and chickpea plants grown in Kibbutz Kfar H'horesh (32°42′7.56″ N 35°16′27.47″ E), respectively. The inflorescences were dried at 23–35 °C for 2 months and then the seeds were separated with a 300-mesh size sieve (50 µm) and stored in the dark at 4 °C until use.

4.2. Mutagenesis

WT F01 chickpea breeding line seeds were used for mutagenesis. Approximately 6000 WT F01 seeds were allowed to swell in water for 10 h and then exposed to the mutation inducer EMS at a concentration of 4% (vol/vol) which, according to the dose-response curve, decreased seed germination by 50%. After shaking at 50 rpm for 10 h, the EMS was removed, and the seeds were washed under running tap water for 14 h. The seeds were dried under airflow for 48 h and delivered to Shorashim Nursery Ltd., Israel, to produce seedlings. The seedlings were planted and grown in a field at the Western Galilee experimental farm, Israel (32°55′ N 35°04′ E), to produce M_2 seeds.

4.3. Screening for Broomrape Resistance

An EMS-mutated population of about 3000 families (each derived from a single M_1 plant) was used to screen for broomrape resistance. Eight M_2 generation seeds from each family were seeded separately in soil containing seeds of *P. aegyptiaca* and *O. crenata* at a concentration of 20 mg seeds per kg of soil (~3000 seeds/kg). After 3 months, plant roots were evaluated for broomrape infection. Families of plants that were free of broomrape were selected for the next screening, leading to identification of the broomrape-resistant mutant CCD7M14.

4.4. Phenotype Determination

4.4.1. Evaluation of Broomrape Resistance

Broomrape-resistance tests were conducted in 2 L pots, each filled with soil mixed with the seeds of *P. aegyptiaca* and *O. crenata* at a concentration of 20 mg seeds per kg soil. Control pots did not contain broomrape seeds. Each pot was planted with one chickpea plant. Organic medium-heavy clay–loam soil collected in Newe Ya'ar Research Center ($32°42'9''$ N, $35°10'9''$ E) was used in all experiments. The plants were grown in nethouse and irrigated and fertilized as needed. The experiments were arranged in a completely randomized design with 10 replications (pots) per treatment. Once a week, the number of broomrape shoots per pot was evaluated. At the end of the experiments, the roots were gently washed out of the pots under tap water and broomrape number and fresh biomass were determined.

4.4.2. Resistance Mechanism Determination

The ability of WT and mutant plants to induce germination of *P. aegyptiaca* and *O. crenata* seeds was tested in GF/A glass microfiber filter paper envelopes [25]. Briefly, *P. aegyptiaca* or *O. crenata* seeds inside the paper envelopes were placed close to the chickpea roots at planting. Seed germination percentage was recorded four weeks after planting using a stereoscopic microscope. Control pots (without plants) were used for spontaneous seed germination determination.

To analyze SLs in root exudates, WT F01 and CCD7M14 plants were grown under hydroponic conditions with feeding solution circulated through activated charcoal [25]. Once a week, the charcoal was washed with water and extracted with acetone. The acetone solutions were combined and evaporated under reduced pressure at 35 °C (Rotavapor, Büchi, Switzerland) from all samples. The residue was dissolved in 200 mL water and the solution was extracted three times with equal volumes of ethyl acetate. The ethyl acetate fractions were combined, washed with 0.2 M K_2HPO_4 (pH 8.3), dried over anhydrous Na_2SO_4, and concentrated under reduced pressure at 35 °C. Dry extracts were stored at 4 °C.

Samples of root exudates were tested for the ability to germinate preconditioned *P. aegyptiaca* seeds according to Yoneyama et al. (2007) [65]. Briefly, dried root exudates were dissolved in methanol up to concentration of 0.2, 2 and 20 μg/mL of which 100 μL was applied to filter paper inside 45-mm diameter Petri dishes. After drying under air flour, 0.2 mL of sterile water was added to the disks to get final concentrations of 0.1, 1, and 10 μg/mL. Disinfected *P. aegyptiaca* seeds were distributed on a 45 mm filter paper disk and kept moistened for 1 week. Then the disks with seeds on them were dried gently on sterile filter paper and transferred to the Petri dishes upon the disks containing root exudates. For the positive control, stimulation with GR24 at a concentration of 10^{-6} M was used. The plates were kept at 25 °C for 10 days, and the *P. aegyptiaca* seed germination was evaluated utilizing of a stereoscopic microscope.

LC-MS/MS analysis of proton adduct ions of SLs was performed with a triple quadrupole/linear ion trap instrument (LIT) (QTRAP5500; AB Sciex) with an electrospray source according to Yoneyama et al., 2007 [65]. All peaks corresponding to strigolactones were confirmed by *P. aegyptiaca* seed-germination assay [25].

4.4.3. Plant Morphology and Pigment Contents

The plants of WT F01 and CCD7M14 were grown in 4 L pots in Newe Ya'ar organic soil. After 14 weeks, the plants were harvested by cutting the stems at the pot's soil surface. First, third and fifth leaves were sampled for determination of total carotenoid, anthocyanin, and chlorophyll a and b contents. The number of primary branches, the number of secondary branches per primary branch, and foliage and root fresh biomass were determined.

Contents of carotenoids and anthocyanin were measured according to Segev et al. [66], and chlorophyll were was measured according to Lichtenthaler [67]. Briefly, chlorophyll and anthocyanin were extracted using methanol and acidic methanol (99% methanol and

1% hydrochloric acid), respectively. The test tubes were incubated at room temperature for two days in the dark. After two days, the solutions were tested in a spectrophotometer at 665, 652, 530, and 470 nm wavelengths. From the data we calculated the relative amounts of total chlorophyll, chlorophyll a, chlorophyll b, total carotenoids, and total anthocyanins according to the following formulas:

$$\text{Chlorophyll A (μg/mL)} = 16.72 \times A_{665} - 9.16 \times A_{652};$$
$$\text{Chlorophyll B (μg/mL)} = 34.09 \times A_{652} - 15.28 \times A_{665};$$
$$\text{Total chlorophyll (a + b) (μg/mL)} = 1.44 \times A_{665} + 24.93 \times A_{652};$$
$$\text{Total carotenoids (μg/mL)} = (1000 \times A_{470} - 1.63 \times \text{Chlorophyll A} - 104.96 \times \text{Chlorophyll B})/221;$$
$$\text{Total anthocyanins (μg/mL)} = (449.1 \times A_{530} + 24.93 \times 2000)/24{,}500.$$

The final results were calculated in μg per 1 g of fresh leaf biomass.

4.5. DNA Extraction and PCR Amplification

Total genomic DNA was extracted from young leaves of 2-week-old M_3 plants homozygous for broomrape resistance. Primer design, PCR amplification, electrophoresis in a 1.0% agarose gel, and sequence analysis of the *CCD7* gene were performed as described by Schreiber et al. [68], with several modifications: annealing was performed at 55 °C for 30 s and synthesis at 72 °C for 60 s. Eight pairs of primers, purchased from Syntezza Bioscience Ltd. (Jerusalem, Israel), were used (Table 6).

Table 6. Primer sets used in this study.

Primer Set	Exon	Forward Primer	Reverse Primer	Product Size (bp)	Sequenced Region (cDNA)
1	1	AGCACATTTTGTTGCCAAGC	TCCTGCTTACATGAAATGCAAACT	1090	1–529
2	1	GAGTACGATCGAAAGACTGACTCG	TCCTGCTTACATGAAATGCAAACT	551	522–776
3	2	TACAAGGTGTACAACATTGAGT	ACTGCCAATTTGTTGGCATTTC	599	777–908
4	3	GAAATGCCAACAAATTGGCAGT	GCATGCTTAAATTTCATTTTGGA	621	909–1043
5	4	TCATGAGGGAGTAAATAATCAACA	TTTAATTCACGTTTTATGTCGGT	623	1044–1316
6	5	AGGGACAAAAATTATCGGCTT	CTTAGGATAAACCACACATAGATAG	361	1317–1404
7	6	CCAATTAAGATGTTCGAGAGCT	ACATGGACAAATCTATAACGACA	747	1405–1710
8	7	AGTAATAGCTAATCAAAACGGGT	TTGGATTTCCAAGAGTCCAAT	686	1711–1872

4.6. Statistical Analysis

All experimental results were subjected to ANOVA utilizing JMP software, version 5.0 (SAS Institute Inc., Cary, NC, USA). Data on seed germination were separated by standard error of the mean (SEM). To meet the assumption of ANOVA, percentage data were arcsine-transformed before analysis. The results on the number of aboveground broomrape shoots were compared by SEM and by least-significant differences (LSD), based on Tukey–Kramer honestly significant difference test ($\alpha = 0.05$). Data on the number and biomass of *P. aegyptiaca* and *O. crenata* attached to chickpea roots after root washing were separated by SEM. The experiments on chickpea lines sensitivity to *P. aegyptiaca* and *O. crenata* were repeated twice. The repeated experiments were compared using Fisher's *t*-test, which showed homogeneity of variances; therefore, the data were combined. The test of the differences in morphology between WT F01 and CCD7M14 and the data of pigment concentration in chickpea leaves was conducted with 5 and 3 replicates, respectively and separated by SEM. The results were analyzed by LS means contrast test ($\alpha = 0.05$).

5. Conclusions

Using EMS mutagenesis, chickpea line CCD7M14 showing high resistance to both *O. crenata* and *P. aegyptiaca* was developed. The resistance mechanism was based on blockage of SL synthesis, probably caused by stop codon formation due to the point mutation in the *CCD7* gene. Root exudates of the mutant did not contain SLs. The mutant plants displayed increased branching and reduced plant height; decreased chlorophyll and carotenoid contents; and increased accumulation of anthocyanin in the leaves compared with the WT.

Supplementary Materials: The following are available online at https://www.mdpi.com/article/10.3390/plants10122552/s1, Figure S1: WT F01 (right) and CCD7M14 (left) plants growing in soil mixed with seeds of *P. aegyptiaca* at a concentration of 20 mg seeds per kg soil. Figure S2: Selected reaction monitoring (SRM) chromatograms of WT F01 (**a**) and CCD7M14 (**b**) root exudates. Determination of SLs was based on the retention time and transition of m/z 345 > 97 for didehydroorobanchol and 347 > 233 for orobanchol and orobanchyl acetate.

Author Contributions: Conceptualization, S.G., J.H. and E.D.; Methodology, J.H. and E.D.; Software, S.G.; Investigation, E.S., O.A.-S., K.Y., X.X. and A.B.; Resources, S.G. and J.H.; Data Curation, E.D.; Writing—Original Draft Preparation, S.G., J.H. and E.D.; Writing—Review and Editing, S.G., J.H., K.Y. and E.D.; Visualization, S.G. and E.S.; Supervision, S.G. and E.D. All authors have read and agreed to the published version of the manuscript.

Funding: This research received no external funding.

Data Availability Statement: Datais contained within the article and Supplementary Material.

Conflicts of Interest: The authors declare no conflict of interest.

References

1. FAOSTAT. Available online: http://faostat.fao.org/default.aspx (accessed on 1 January 2016).
2. Galili, S.; Hovav, R.; Dor, E.; Hershenhorn, J.; Harel, A.; Amir-Segev, O.; Bellalou, A.; Badani, H.; Smirnov, E.; Achdari, G. The history of chickpea cultivation and breeding in Israel. *Isr. J. Plant Sci.* **2018**, *65*, 186–194. [CrossRef]
3. Dor, E.; Smirnov, E.; Galili, S.; Guy, A.; Hershenhorn, J. Characterization of the novel tomato mutant HRT, resistant to acetolactate synthase–inhibiting herbicides. *Weed Sci.* **2016**, *64*, 348–360. [CrossRef]
4. Dor, E.; Galili, S.; Smirnov, E.; Hacham, Y.; Amir, R.; Hershenhorn, J. The effects of herbicides targeting aromatic and branched chain amino acid biosynthesis support the presence of functional pathways in broomrape. *Front. Plant Sci.* **2017**, *8*, 707. [CrossRef] [PubMed]
5. Venezian, A.; Dor, E.; Achdari, G.; Plakhine, D.; Smirnov, E.; Hershenhorn, J. The influence of the plant growth regulator maleic hydrazide on Egyptian broomrape early developmental stages and its control efficacy in tomato under greenhouse and field conditions. *Front. Plant Sci.* **2017**, *8*, 691. [CrossRef] [PubMed]
6. Parker, C.; Riches, C.R. *Parasitic Weeds of the World: Biology and Control*; CAB International: Wallingford, UK, 1993.
7. Joel, D.M.; Hershenhorn, J.; Eizenberg, H.; Aly, R.; Ejeta, G.; Rich, P.J.; Ransom, J.K.; Sauerborn, J.; Rubiales, D. Biology and management of weedy root parasites. In *Horticultural Reviews*; John Wiley & Sons: Hoboken, NJ, USA, 2007; Volume 33. [CrossRef]
8. Joel, D.M.; Kleifeld, Y.; Losner-Goshen, D.; Herzlinger, G.; Gressel, J. Transgenic crops against parasites. *Nature* **1995**, *374*, 220–221. [CrossRef]
9. Yokota, T.; Sakai, H.; Okuno, K.; Yoneyama, K.; Takeuchi, Y. Alectrol and orobanchol, germination stimulants for *Orobanche minor*, from its host red clover. *Phytochemistry* **1998**, *49*, 1967–1973. [CrossRef]
10. Xie, X.; Yoneyama, K.; Yoneyama, K. The strigolactone story. *Annu. Rev. Phytopathol.* **2010**, *48*, 93–117. [CrossRef] [PubMed]
11. Boyer, F.-D.; de Saint Germain, A.; Pillot, J.-P.; Pouvreau, J.-B.; Chen, V.X.; Ramos, S.; Stévenin, A.; Simier, P.; Delavault, P.; Beau, J.-M.; et al. Structure-activity relationship studies of strigolactone-related molecules for branching inhibition in garden pea: Molecule design for shoot branching. *Plant Physiol.* **2012**, *159*, 1524–1544. [CrossRef] [PubMed]
12. Sorefan, K.; Booker, J.; Haurogné, K.; Goussot, M.; Bainbridge, K.; Foo, E.; Chatfield, S.; Ward, S.; Beveridge, C.; Rameau, C.; et al. MAX4 and RMS1 are orthologous dioxygenase-like genes that regulate shoot branching in Arabidopsis and pea. *Genes Dev.* **2003**, *17*, 1469–1474. [CrossRef] [PubMed]
13. Booker, J.; Auldridge, M.; Wills, S.; McCarty, D.; Klee, H.; Leyser, O. MAX3/CCD7 is a carotenoid cleavage dioxygenase required for the synthesis of a novel plant signaling molecule. *Curr. Biol.* **2004**, *14*, 1232–1238. [CrossRef] [PubMed]
14. Matusova, R.; Rani, K.; Verstappen, F.W.A.; Franssen, M.C.R.; Beale, M.H.; Bouwmeester, H.J. The strigolactone germination stimulants of the plant-parasitic *Striga* and *Orobanche* spp. are derived from the carotenoid pathway. *Plant Physiol.* **2005**, *139*, 920–934. [CrossRef]
15. Alder, A.; Jamil, M.; Marzorati, M.; Bruno, M.; Vermathen, M.; Bigler, P.; Ghisla, S.; Bouwmeester, H.J.; Beyer, P.; Al-Babili, S. The path from β-carotene to carlactone, a strigolactone-like plant hormone. *Science* **2012**, *335*, 1348–1351. [CrossRef] [PubMed]
16. Brewer, P.B.; Yoneyama, K.; Filardo, F.; Meyers, E.; Scaffidi, A.; Frickey, T.; Akiyama, K.; Seto, Y.; Dun, E.A.; Cremer, J.E.; et al. Lateral branching oxidorreductase acts in the final stages of strigolactone biosynthesis in Arabidopsis. *Proc. Natl. Acad. Sci. USA* **2016**, *113*, 6301–6306. [CrossRef] [PubMed]

17. Seto, Y.; Sado, A.; Asami, K.; Hanada, A.; Umehara, M.; Akiyama, K.; Yamaguchi, S. Carlactone is an endogenous biosynthetic precursor for strigolactones. *Proc. Natl. Acad. Sci. USA* **2014**, *111*, 1640–1645. [CrossRef] [PubMed]

18. Zhang, Y.; Van Dijk, A.D.J.; Scaffidi, A.; Flematti, G.R.; Hofmann, M.; Charnikhova, T.; Verstappen, F.W.A.; Hepworth, J.; Van Der Krol, S.; Leyser, O.; et al. Rice cytochrome P450 MAX1 homologs catalyze distinct steps in strigolactone biosynthesis. *Nat. Chem. Biol.* **2014**, *10*, 1028–1033. [CrossRef]

19. Abe, S.; Sado, A.; Tanaka, K.; Kisugi, T.; Asami, K.; Ota, S.; Kim, H.I.; Yoneyama, K.; Xie, X.; Ohnishi, T.; et al. Carlactone is converted to carlactonoic acid by MAX1 in Arabidopsis and its methyl ester can directly interact with AtD14 in vitro. *Proc. Natl. Acad. Sci. USA* **2014**, *111*, 18084–18089. [CrossRef]

20. Al-Babili, S.; Bouwmeester, H.J. Strigolactones, a novel carotenoid-derived plant hormone. *Annu. Rev. Plant Biol.* **2015**, *66*, 161–186. [CrossRef]

21. Kretzschmar, T.; Kohlen, W.; Sasse, J.; Borghi, L.; Schlegel, M.; Bachelier, J.B.; Reinhardt, D.; Bours, R.; Bouwmeester, H.J.; Martinoia, E. A petunia ABC protein controls strigolactone-dependent symbiotic signalling and branching. *Nature* **2012**, *483*, 341–344. [CrossRef]

22. Sasse, J.; Simon, S.; Gübeli, C.; Liu, G.-W.; Cheng, X.; Friml, J.; Bouwmeester, H.J.; Martinoia, E.; Borghi, L. Asymmetric localizations of the ABC transporter PaPDR1 trace paths of directional strigolactone transport. *Curr. Biol.* **2015**, *25*, 647–655. [CrossRef]

23. Delavault, P.; Montiel, G.; Brun, G.; Pouvreau, J.-B.; Thoiron, S.; Simier, P. Communication between host plants and parasitic plants. *Adv. Bot. Res.* **2017**, *82*, 55–82.

24. Dor, E.; Alperin, B.; Wininger, S.; Ben-Dor, B.; Somvanshi, V.S.; Koltai, H.; Kapulnik, Y.; Hershenhorn, J. Characterization of a novel tomato mutant resistant to *Orobanche* and *Phelipanche* spp. weedy parasites. *Euphytica* **2010**, *171*, 371–380. [CrossRef]

25. Dor, E.; Yoneyama, K.; Wininger, S.; Kapulnik, Y.; Yoneyama, K.; Koltai, H.; Xie, X.; Hershenhorn, J. Strigolactone deficiency confers resistance in tomato line SL-ORT1 to the parasitic weeds *Phelipanche* and *Orobanche* spp. *Phytopathology* **2011**, *101*, 213–222. [CrossRef] [PubMed]

26. Ejeta, G. Breeding for *Striga* resistance in sorghum: Exploitation of an intricate host–parasite biology. *Crop Sci.* **2007**, *47*, S216–S227. [CrossRef]

27. Jamil, M.; Rodenburg, J.; Charnikhova, T.; Bouwmeester, H.J. Pre-attachment *Striga hermonthica* resistance of new rice for Africa (NERICA) cultivars based on low strigolactone production. *New Phytol.* **2011**, *192*, 964–975. [CrossRef] [PubMed]

28. Fondevilla, S.; Fernández-Aparicio, M.; Satovic, Z.; Emeran, A.A.; Torres, A.M.; Moreno, M.T.; Rubiales, D. Identification of quantitative trait loci for specific mechanisms of resistance to *Orobanche crenata* Forsk. in pea (*Pisum sativum* L.). *Mol. Breed.* **2010**, *25*, 259–272. [CrossRef]

29. Fernandez-Aparicio, M.; Kisugi, T.; Xie, X.; Rubiales, D.; Yoneyama, K. Low strigolactone root exudation: A novel mechanism of broomrape (*Orobanche* and *Phelipanche* spp.) resistance available for faba bean breeding. *J. Agric. Food Chem.* **2014**, *62*, 7063–7071. [CrossRef]

30. Gomez-Roldan, V.; Fermas, S.; Brewer, P.B.; Puech-Pagès, V.; Dun, E.A.; Pillot, J.-P.; Letisse, F.; Matusova, R.; Danoun, S.; Portais, J.-C.; et al. Strigolactone inhibition of shoot branching. *Nature* **2008**, *455*, 189–194. [CrossRef]

31. Umehara, M.; Hanada, A.; Yoshida, S.; Akiyama, K.; Arite, T.; Takeda-Kamiya, N.; Magome, H.; Kamiya, Y.; Shirasu, K.; Yoneyama, K.; et al. Inhibition of shoot branching by new terpenoid plant hormones. *Nature* **2008**, *455*, 195–200. [CrossRef]

32. Ruyter-Spira, C.; Kohlen, W.; Charnikhova, T.; van Zeijl, A.; van Bezouwen, L.; de Ruijter, N.; Cardoso, C.; Lopez-Raez, J.A.; Matusova, R.; Bours, R.; et al. Physiological effects of the synthetic strigolactone analog GR24 on root system architecture in Arabidopsis: Another belowground role for strigolactones? *Plant Physiol.* **2011**, *155*, 721–734. [CrossRef]

33. Kapulnik, Y.; Delaux, P.-M.; Resnick, N.; Mayzlish-Gati, E.; Wininger, S.; Bhattacharya, C.; Séjalon-Delmas, N.; Combier, J.-P.; Bécard, G.; Belausov, E.; et al. Strigolactones affect lateral root formation and root-hair elongation in Arabidopsis. *Planta* **2011**, *233*, 209–216. [CrossRef]

34. Koltai, H.; Dor, E.; Hershenhorn, J.; Joel, D.M.; Weininger, S.; Lekalla, S.; Shealtiel, H.; Bhattacharya, C.; Eliahu, E.; Resnick, N.; et al. Strigolactones' effect on root growth and root-hair elongation may be mediated by auxin-efflux carriers. *J. Plant Growth Regul.* **2010**, *29*, 129–136. [CrossRef]

35. Koltai, H.; LekKala, S.P.; Bhattacharya, C.; Mayzlish-Gati, E.; Resnick, N.; Wininger, S.; Dor, E.; Yoneyama, K.; Yoneyama, K.; Hershenhorn, J.; et al. A tomato strigolactone-impaired mutant displays aberrant shoot morphology and plant interactions. *J. Exp. Bot.* **2010**, *61*, 1739–1749. [CrossRef] [PubMed]

36. Bai, J.; Wei, Q.; Shu, J.; Gan, Z.; Li, B.; Yan, D.; Huang, Z.; Guo, Y.; Wang, X.; Zhang, L. Exploration of resistance to *Phelipanche aegyptiaca* in tomato. *Pest Manag. Sci.* **2020**, *76*, 3806–3821. [CrossRef]

37. Bari, V.K.; Nassar, J.A.; Kheredin, S.M.; Gal-On, A.; Ron, M.; Britt, A.; Steele, D.; Yoder, J.; Aly, R. CRISPR/Cas9-mediated mutagenesis of Carotenoid Cleavage Dioxygenase 8 in tomato provides resistance against the parasitic weed *Phelipanche aegyptiaca*. *Sci. Rep.* **2019**, *9*, 11438. [CrossRef] [PubMed]

38. Bari, V.K.; Nassar, J.A.; Meir, A.; Aly, R. Targeted mutagenesis of two homologous ATP-binding cassette subfamily G (ABCG) genes in tomato confers resistance to parasitic weed *Phelipanche aegyptiaca*. *J. Plant Res.* **2021**, *134*, 585–597. [CrossRef]

39. Pavan, S.; Schiavulli, A.; Marcotrigiano, A.R.; Bardaro, N.; Bracuto, V.; Ricciardi, F.; Charnikhova, T.; Lotti, C.; Bouwmeester, H.J.; Ricciardi, L. Characterization of low-strigolactone germplasm in pea (*Pisum sativum* L.) resistant to crenate broomrape (*Orobanche crenata* Forsk.). *Mol. Plant-Microbe Interact.* **2016**, *29*, 743–749. [CrossRef]

40. Fernández-Aparicio, M.; Moral, A.; Kharrat, M.; Rubiales, D. Resistance against broomrapes (*Orobanche* and *Phelipanche* spp.) in faba bean (*Vicia faba*) based in low induction of broomrape seed germination. *Euphytica* **2012**, *186*, 897–905. [CrossRef]

41. Vallabhaneni, R.; Bradbury, L.M.T.; Wurtzel, E.T. The carotenoid dioxygenase gene family in maize, sorghum, and rice. *Arch. Biochem. Biophys.* **2010**, *504*, 104–111. [CrossRef]

42. Yoneyama, K.; Xie, X.; Yoneyama, K.; Takeuchi, Y. Strigolactones: Structures and biological activities. *Pest Manag. Sci.* **2009**, *65*, 467–470. [CrossRef]

43. Wang, Y.; Bouwmeester, H.J. Structural diversity in the strigolactones. *J. Exp. Bot.* **2018**, *69*, 2219–2230. [CrossRef]

44. Xie, X.; Kusumoto, D.; Takeuchi, Y.; Yoneyama, K.; Yamada, Y.; Yoneyama, K. 2′-Epi-orobanchol and solanacol, two unique strigolactones, germination stimulants for root parasitic weeds, produced by tobacco. *J. Agric. Food Chem.* **2007**, *55*, 8067–8072. [CrossRef] [PubMed]

45. Tokunaga, T.; Hayashi, H.; Akiyama, K. Medicaol, a strigolactone identified as a putative didehydro-orobancholisomer, from *Medicago truncatula*. *Phytochemistry* **2015**, *111*, 91–97. [CrossRef] [PubMed]

46. Xie, X.; Yoneyama, K.; Kusumoto, D.; Yamada, Y.; Yokota, T.; Takeuchi, Y.; Yoneyama, K. Isolation and identification of alectrol as (+)-orobanchyl acetate, a germination stimulant for root parasitic plants. *Phytochemistry* **2008**, *69*, 427–431. [CrossRef]

47. Xie, X.; Yoneyama, K.; Kisugi, T.; Uchida, K.; Ito, S.; Akiyama, K.; Hayashi, H.; Yokota, T.; Nomura, T.; Yoneyama, K. Confirming stereochemical structures of strigolactones produced by rice and tobacco. *Mol. Plant* **2013**, *6*, 153–163. [CrossRef] [PubMed]

48. Yoneyama, K.; Xie, X.; Kisugi, T.; Nomura, T.; Sekimoto, H.; Yokota, T.; Yoneyama, K. Characterization of strigolactones exuded by Asteraceae plants. *Plant Growth Regul.* **2011**, *65*, 495–504. [CrossRef]

49. Trabelsi, I.; Yoneyama, K.; Abbes, Z.; Amri, M.; Xie, X.; Kisugi, T.; Kim, H.I.; Kharrat, M.; Yoneyama, K. Characterization of strigolactones produced by *Orobanche foetida* and *Orobanche crenata* resistant faba bean (*Vicia faba* L.) genotypes and effects of phosphorous, nitrogen, and potassium deficiencies on strigolactone production. *South Afr. J. Bot.* **2017**, *108*, 15–22. [CrossRef]

50. Vogel, J.T.; Walter, M.H.; Giavalisco, P.; Lytovchenko, A.; Kohlen, W.; Charnikhova, T.; Simkin, A.J.; Goulet, C.; Strack, D.; Bouwmeester, H.J.; et al. SlCCD7 controls strigolactone biosynthesis, shoot branching and mycorrhiza-induced apocarotenoid formation in tomato. *Plant J.* **2010**, *61*, 300–311. [CrossRef]

51. De Saint Germain, A.; Ligerot, Y.; Dun, E.A.; Pillot, J.-P.; Ross, J.J.; Beveridge, C.A.; Rameau, C. Strigolactones stimulate internode elongation independently of gibberellins. *Plant Physiol.* **2013**, *163*, 1012–1025. [CrossRef]

52. Snowden, K.C.; Simkin, A.J.; Janssen, B.J.; Templeton, K.R.; Loucas, H.M.; Simons, J.L.; Karunairetnam, S.; Gleave, A.P.; Clark, D.G.; Klee, H.J. The decreased apical dominance1/Petunia hybrida Carotenoid Cleavage Dioxygenase 8 gene affects branch production and plays a role in leaf senescence, root growth, and flower development. *Plant Cell* **2005**, *17*, 746–759. [CrossRef]

53. Muhr, M.; Prüfer, N.; Paulat, M.; Teichmann, T. Knockdown of strigolactone biosynthesis genes in *Populus* affects branched 1 expression and shoot architecture. *New Phytol.* **2016**, *212*, 613–626. [CrossRef]

54. Ongaro, V.; Leyser, O. Hormonal control of shoot branching. *J. Exp. Bot.* **2008**, *59*, 67–74. [CrossRef] [PubMed]

55. Wakabayashi, T.; Hamana, M.; Mori, A.; Akiyama, R.; Ueno, K.; Osakabe, K.; Osakabe, Y.; Suzuki, H.; Takikawa, H.; Mizutani, M.; et al. Direct conversion of carlactonoic acid to orobanchol by cytochrome P450 CYP722C in strigolactone biosynthesis. *Sci. Adv.* **2019**, *5*, eaax9067. [CrossRef]

56. Min, Z.; Li, R.; Chen, L.; Zhang, Y.; Li, Z.; Liu, M.; Ju, Y.; Fang, Y. Alleviation of drought stress in grapevine by foliar-applied strigolactones. *Plant Physiol. Biochem.* **2019**, *135*, 99–110. [CrossRef] [PubMed]

57. Hu, Q.; Ding, F.; Li, M.; Zhang, X.; Zhang, S.; Huang, B. Strigolactone and ethylene inhibitor suppressing dark-induced leaf senescence in perennial ryegrass involving transcriptional downregulation of chlorophyll degradation. *J. Am. Soc. Hortic. Sci.* **2021**, *146*, 79–86. [CrossRef]

58. Qiu, C.-W.; Zhang, C.; Wang, N.-H.; Mao, W.; Wu, F. Strigolactone GR24 improves cadmium tolerance by regulating cadmium uptake, nitric oxide signaling and antioxidant metabolism in barley (*Hordeum vulgare* L.). *Environ. Pollut.* **2021**, *273*, 116486. [CrossRef]

59. Zheng, X.; Li, Y.; Xi, X.; Ma, C.; Sun, Z.; Yang, X.; Li, X.; Tian, Y.; Wang, C. Exogenous strigolactones alleviate KCl stress by regulating photosynthesis, ROS migration and ion transport in *Malus hupehensis* Rehd. *Plant Physiol. Biochem.* **2021**, *159*, 113–122. [CrossRef]

60. Sarwar, Y.; Shahbaz, M. Modulation in growth, photosynthetic pigments, gas exchange attributes and inorganic ions in sunflower (*Helianthus annuus* L.) by strigolactones (GR24) achene priming under saline conditions. *Pak. J. Bot.* **2020**, *52*, 23–31. [CrossRef]

61. Ferrero, M.; Pagliarani, C.; Novák, O.; Ferrandino, A.; Cardinale, F.; Visentin, I.; Schubert, A. Exogenous strigolactone interacts with abscisic acid-mediated accumulation of anthocyanins in grapevine berries. *J. Exp. Bot.* **2018**, *69*, 2391–2401. [CrossRef]

62. Li, S.; Joo, Y.; Cao, D.; Li, R.; Lee, G.; Halitschke, R.; Baldwin, G.; Baldwin, I.T.; Wang, M. Strigolactone signaling regulates specialized metabolism in tobacco stems and interactions with stem-feeding herbivores. *PLoS Biol.* **2020**, *18*, e3000830. [CrossRef]

63. Li, W.; Nguyen, K.H.; Tran, C.D.; Watanabe, Y.; Tian, C.; Yin, X.; Li, K.; Yang, Y.; Guo, J.; Miao, Y.; et al. Negative roles of strigolactone-related SMXL6, 7 and 8 proteins in drought resistance in Arabidopsis. *Biomolecules* **2020**, *10*, 607. [CrossRef]

64. Wang, L.; Yu, H.; Guo, H.; Lin, T.; Kou, L.; Wang, A.; Shao, N.; Ma, H.; Xiong, G. Transcriptional regulation of strigolactone signalling in Arabidopsis. *Nature* **2020**, *583*, 277–281. [CrossRef]

65. Yoneyama, K.; Yoneyama, K.; Takeuchi, Y.; Sekimoto, H. Phosphorus deficiency in red clover promotes exudation of orobanchol, the signal for mycorrhizal symbionts and germination stimulant for root parasites. *Planta* **2007**, *225*, 1031–1038. [CrossRef] [PubMed]

66. Segev, A.; Badani, H.; Kapulnik, Y.; Shomer, I.; Oren-Shamir, M.; Galili, S. Determination of polyphenols, flavonoids, and antioxidant capacity in colored chickpea (*Cicer arietinum* L.). *J. Food Sci.* **2010**, *75*, S115–S119. [CrossRef]
67. Lichtenthaler, H.K. Chlorophylls and carotenoids: Pigments of photosynthetic biomembranes. *Methods Enzymol.* **1987**, *148*, 350–382.
68. Schreiber, G.; Reuveni, M.; Evenor, D.; Oren-Shamir, M.; Ovadia, R.; Sapir-Mir, M.; Bootbool-Man, A.; Nahon, S.; Shlomo, H.; Chen, L.; et al. Anthocyanin1 from *Solanum chilense* is more efficient in accumulating anthocyanin metabolites than its *Solanum lycopersicum* counterpart in association with the Anthocyanin Fruit phenotype of tomato. *TAG* **2012**, *124*, 295–307. [CrossRef] [PubMed]

Article

Trophic Transfer without Biomagnification of Cadmium in a Soybean-Dodder Parasitic System

Bin J. W. Chen *, Jing Xu and Xinyu Wang

College of Biology and the Environment, Nanjing Forestry University, Nanjing 210037, China; xujing0323@outlook.com (J.X.); Wangxinyu1007@outlook.com (X.W.)
* Correspondence: bin.chen@njfu.edu.cn

Abstract: Cadmium (Cd) is among the most available and most toxic heavy metals taken up by plants from soil. Compared to the classic plant-animal food chains, the host-parasitic plant food chains have, thus far, been largely overlooked in the studies of Cd trophic transfer. To investigate the pattern of Cd transfer during the infection of parasitic plants on Cd-contaminated hosts, we conducted a controlled experiment that grew soybeans parasitized by Chinese dodders (*Cuscuta chinensis*) in soil with different levels of Cd treatment, and examined the concentration, accumulation, allocation and transfer coefficients of Cd within this parasitic system. Results showed that among all components, dodders accounted for more than 40% biomass of the whole system but had the lowest Cd concentration and accumulated the least amount of Cd. The transfer coefficient of Cd between soybean stems and dodders was much lower than 1, and was also significantly lower than that between soybean stems and soybean leaves. All these features were continuously strengthened with the increase of Cd treatment levels. The results suggested no evidence of Cd biomagnification in dodders parasitizing Cd-contaminated hosts, and implied that the Cd transfer from hosts to dodders may be a selective process.

Keywords: *Cuscuta*; food chain; feeding mode; heavy metal; holoparasite; host; parasitic plants

Citation: Chen, B.J.W.; Xu, J.; Wang, X. Trophic Transfer without Biomagnification of Cadmium in a Soybean-Dodder Parasitic System. *Plants* **2021**, *10*, 2690. https://doi.org/10.3390/plants10122690

Academic Editors: Evgenia Dor and Yaakov Goldwasser

Received: 3 November 2021
Accepted: 5 December 2021
Published: 7 December 2021

Publisher's Note: MDPI stays neutral with regard to jurisdictional claims in published maps and institutional affiliations.

1. Introduction

Along with worldwide industrialization over the last century, environmental pollution has become an important global issue [1]. Heavy metals, i.e., (semi-)metallic elements with an atomic density > 5 g·cm^{-3} [2], have been considered as one of the major types of pollutants [3,4]. Nowadays, the soil has been heavily contaminated by heavy metals, generating serious threats to food safety [5,6] and human health [2,7]. Among various heavy metals, cadmium (Cd) is a non-essential element and can be naturally found in the earth's crust at low concentrations [7]. In addition to some natural processes (e.g., rock weathering and volcanic eruptions), the major sources of Cd contamination in topsoil originate from anthropogenic activities, including phosphate fertilizer applications, industrial waste disposal, fossil fuel combustions, and sewage sludge amendments [4,8,9]. Owing to its relative mobility in soil, Cd is among the most available heavy metals for plant uptake [6,8]. Meanwhile, due to chemical similarities to the divalent ions of some essential metals (e.g., zinc, iron and calcium), Cd ions in soil can easily enter root cells via less-specialized transporters and channels of those ions on the plasma membranes [10]. Cd is also among the most toxic metals to plants [11].

The transfer and accumulation of Cd through food chains have received great attention for more than half a century [9]. There is ample evidence showing that Cd can be biomagnified (i.e., more concentrated) through the trophic levels of food chains in terrestrial ecosystems. For instance, remarkably higher Cd concentrations in the viscera of herbivores and predators than the concentrations in their diets were observed in a Cd-contaminated semi-natural grassland [12]. Approximately 3% of Cd in soil can be transferred to human bodies via the consumption of rice grown in Cd-contaminated farmlands [13].

Parasites are defined as a group of organisms drawing nourishment from a host with only harmful but not immediate lethal effects on the host [14]. Some higher plants have evolved from autotrophic to hemiparasitic or even holoparasitic species. The former (also called 'facultative parasite') is still capable of photosynthesis thus can survive independently of hosts; while the latter (also called 'obligate parasite') has lost photosynthetic function thus fully relying on parasitism to hosts [15]. Among various holoparasites, *Cuscuta* sp. (Convolvulaceae), a.k.a. 'dodders', is a genus of rootless, leafless and string-like stem-parasites that develop connections to the shoots of host plants [16], and are recognized as worldwide agricultural weeds [17]. Newly germinated *Cuscuta* seedlings grow upward and rotate in the air until touching a point for attachment. Once attached and coiled around the stems or leaves of hosts, a special structure called 'haustorium' starts to develop at the contacting point and produce searching hyphae penetrating host tissues [18]. When arriving to the vascular bundles of hosts, the terminal cells of hyphae differentiate and respectively connect to the phloem and xylem of hosts [18,19]. After the establishment of these connections, *Cuscuta* becomes a super sink that compete for water, minerals and photosynthetic assimilates against other sink organs of host plants [20]. The processes of host searching and haustorium induction involve a detection of light quality signaled from host plants [19]. So far, the biological and ecological research of *Cuscuta* mainly focus on their interactions with hosts in the aspects of host selection [21], evolution and development of haustorium [22,23], impacts on host growth [24,25], and exchanges of substances and signals [18,26,27]. Meanwhile, studies of heavy metal stress (especially Cd stress) on the growth of *Cuscuta* are still scarce [28], and most of the published works limited their scopes to the physiological responses and detoxification mechanisms in *Cuscuta* [20,29].

Compared with the path from host plants to animal herbivores, the path from host plants to parasitic plants has received much less attention in the research of trophic transfer and biomagnification of heavy metals [28]. Among various parasitic plants, *Cuscuta* is believed as an ideal model species for studying parasitic trophic transfer of heavy metals in terrestrial ecosystems, since the rootless feature can guarantee that all heavy metals in *Cuscuta* are transferred from hosts without any direct uptake from soil [21]. However, so far to our knowledge, there is no experiment undertaken to investigate the transfer and accumulation of heavy metals, especially Cd, between *Cuscuta* and its hosts grown in contaminated soil. To investigate the transfer pattern and to test the occurrence of biomagnification of Cd in host-*Cuscuta* parasitic systems, we conducted a controlled greenhouse experiment using *C. chinensis* ('dodder' for simplicity, hereafter) as the parasite and soybean (*Glycine max*) as the host grown in soil with a series of Cd amendment levels. We examined the concentration, accumulation and allocation of Cd in various components (i.e., roots, stems and leaves of soybean, as well as dodders) of this soybean-dodder parasitic system, and also evaluated the transfer efficiency of Cd within this system.

2. Results

2.1. Biomass

Cd treatment adversely affected the biomass of all components of the soybean-dodder parasitic system; however, significant reductions in biomass, as compared to that in T0, only occurred in dodders when the level of Cd treatment reached to T4 (Figure 1A). When we focused on the total mass of soybean (i.e., the sum of root, stem and leaf mass), it also tended to continuously decline with the increase of Cd treatment levels. Like the response of dodder mass, a significant reduction in total mass of soybean, as compared to that in T0, was only observed in T4 (Figure 2).

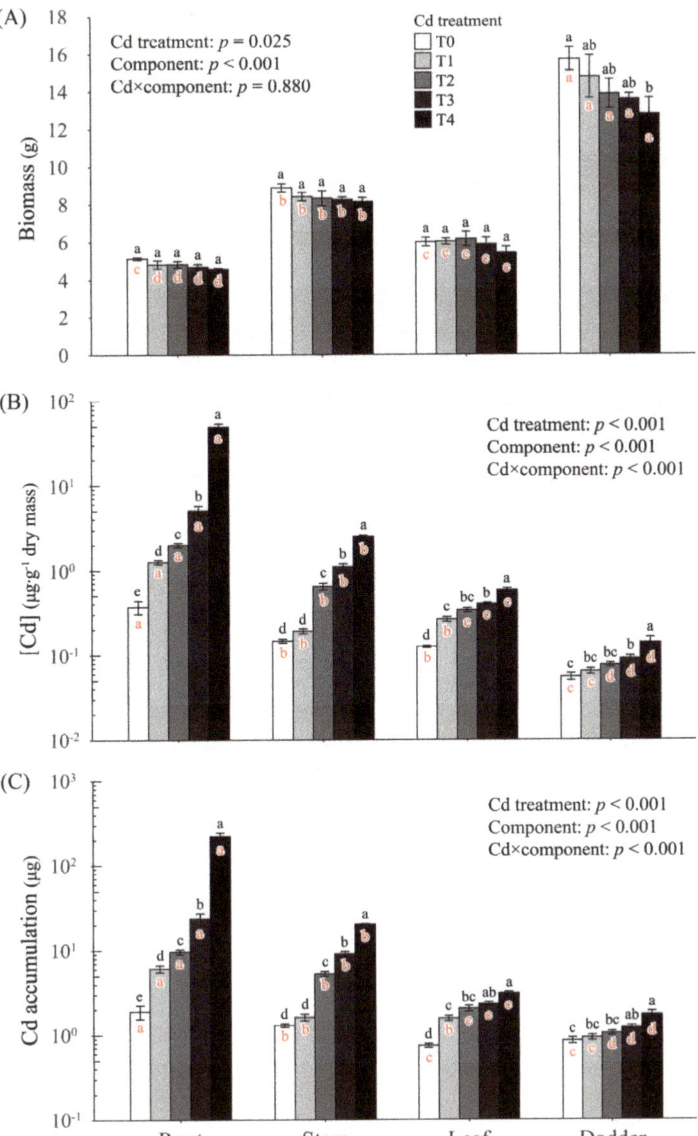

Figure 1. The effects of cadmium (Cd) treatment and component type on the (**A**) biomass, (**B**) Cd concentration ([Cd]) and (**C**) Cd accumulation of various components within the soybean-dodder parasitic systems. The levels of Cd treatment were applied as Hoagland solution (50% strength) amended with 0, 1, 10, 100, and 1000 mg·L^{-1} CdCl$_2$, which are respectively marked as T0, T1, T2, T3, and T4. The analyses were performed using nested two-way ANOVAs with Cd treatment, component type and their interaction term as fixed factors, and pot replicate as a random factor, followed by Tukey's post hoc tests. The results are presented here as p values of fixed factors, which were calculated based on type-III analysis-of-variance. Different black letters within one component type indicate significant differences between Cd treatment levels of that component. Different red letters within one Cd treatment level indicate significant differences between components under that treatment level. The error bars denote 1 SE of the mean.

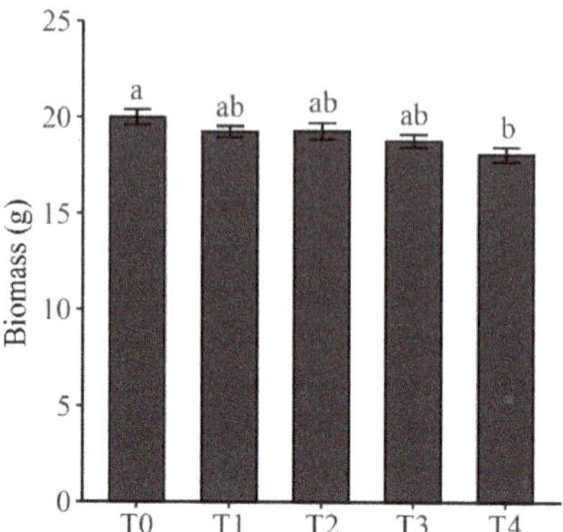

Figure 2. Total mass of soybean plants under different levels of cadmium (Cd) treatment. The levels of Cd treatment were applied as Hoagland solution (50% strength) amended with 0, 1, 10, 100, and 1000 mg·L^{-1} CdCl$_2$, which are respectively marked as T0, T1, T2, T3, and T4. Different letters indicate significant differences between groups. The error bars denote 1 SE of the mean.

2.2. Cd Concentration

With the increase of Cd treatment levels, Cd concentration ([Cd]) in all components also continuously increased. However, compared to [Cd] in T0, the first significant elevation was observed in T1 for roots and leaves, in T2 for stems, and in T3 for dodders (Figure 1B). Within this parasitic system, there was a general pattern that root [Cd] was always the highest, dodder [Cd] was always the lowest, and stem [Cd] and leaf [Cd] were always the intermediate (Figure 1B). However, the rank of [Cd] between stems and leaves appeared dose-dependent in response to Cd treatment. That is, stem [Cd] was similar as leaf [Cd] in T0 and T1 but became significantly higher than leaf [Cd] in higher treatment levels (Figure 1B). Both leaf [Cd] and dodder [Cd] were significantly positively correlated with stem [Cd] (t = 8.836, d.f. = 18, $p < 0.001$ for leaf; t = 7.029, d.f. = 18, $p < 0.001$ for dodder). However, for a given stem [Cd], leaf [Cd] was always higher than dodder [Cd] within the parasitic system; the extent of this difference enlarged with the increase of stem [Cd] (Figure 3).

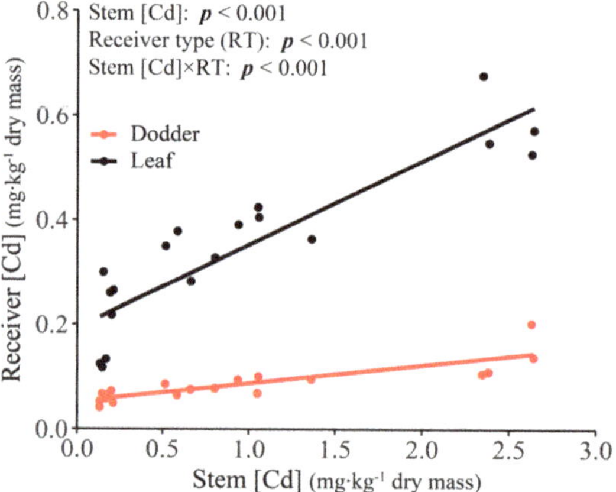

Figure 3. The correlations of cadmium concentration ([Cd]) between soybean stem (source) and its receiver (sink) components (i.e., soybean leaf and dodder). In the analysis (i.e., nested ANCOVA), stem [Cd] covariate, receiver component type and their interaction term were the fixed factors, and pot replicate was the random factor. The results are presented here as *p* values of fixed factors, which were calculated based on type-III analysis-of-variance.

2.3. Cd Transfer Coefficient

Transfer coefficient was defined as the ratio of concentration between sink and source components within the body of the same plant (or animal), or from different trophic levels. The transfer coefficient of Cd of various source-sink paths within soybean plants were almost always lower than 1 and generally declined with the increase of Cd treatment levels (Table 1). However, there were some exceptions. That is, the transfer coefficient of stem-leaf path in T1 was higher than 1 and was significantly higher but not lower than that in T0; and the coefficient of root-stem path in T1 was significantly lower but not higher than that in T2 and T3 (Table 1). Regardless of the levels of Cd treatment, Cd transfer coefficient of stem-leaf path was always significantly higher than that of root-stem path (Table 1).

Table 1. Cadmium (Cd) transfer coefficients of various paths within the soybean-dodder parasitic system under different levels of Cd treatment.

Cd Treatment	Root-Stem	Stem-Leaf	Stem-Dodder
T0	$0.44^a{}_b$ (0.08)	$0.85^b{}_a$ (0.04)	$0.38^a{}_b$ (0.03)
T1	$0.15^{cd}{}_c$ (0.01)	$1.40^a{}_a$ (0.19)	$0.34^{ab}{}_b$ (0.04)
T2	$0.32^{ab}{}_b$ (0.03)	$0.54^c{}_a$ (0.07)	$0.12^b{}_c$ (0.02)
T3	$0.23^{bc}{}_b$ (0.02)	$0.37^{cd}{}_a$ (0.03)	$0.08^c{}_c$ (0.01)
T4	$0.05^d{}_b$ (0.01)	$0.23^d{}_a$ (0.02)	$0.06^c{}_b$ (0.01)

The levels of Cd treatment were applied as Hoagland solution (50% strength) amended with 0, 1, 10, 100, and 1000 mg·L^{-1} CdCl$_2$, which are respectively marked as T0, T1, T2, T3, and T4. Different superscript black letters within one column indicate significant differences between different Cd treatment levels of the same path, while different subscript red letters within one row denote significant differences between different paths under the same Cd treatment level. Values in brackets denote 1 SE of the mean.

When the transfer occurs between different trophic levels, a coefficient with value higher than 1 is a clear sign of biomagnification [30]. Clearly, the transfer coefficient of Cd of the stem-dodder path was always much lower than 1 and declined with the increase of Cd treatment levels without any exception (Table 1). In addition, Cd transfer coefficient of stem-dodder path was always significantly lower than that of stem-leaf path, regardless of Cd treatment levels (Table 1).

2.4. Cd Accumulation

The accumulation of Cd in a component was defined as the absolute amount of Cd in the component. The responses of Cd accumulation in the parasitic system were similar as the responses of [Cd] in the system. For all components, their Cd accumulations continuously increased with the increase of treatment levels. Compared to the accumulation of Cd in T0, the first significant increase was found in T1 for roots and leaves, in T2 for stems, and T3 for dodders (Figure 1C). The rank of Cd accumulation among components was root > stem = leaf > dodder in T0 and T1 but changed to root > stem > leaf > dodder in higher levels of Cd treatment (Figure 1C).

2.5. Allocations of Biomass and Cd

Allocation here was defined as the biomass or Cd accumulation of a component in proportion to the total amount of biomass or Cd of the whole soybean-dodder system. Cd treatment had no effect on biomass allocation pattern of the parasitic system (F = 0.149, p = 0.963). Biomass of dodders always accounted for more than 40% biomass of the whole parasitic system; and the rank of biomass allocation among components was always: dodder > stem > leaf > root (Figure 4A).

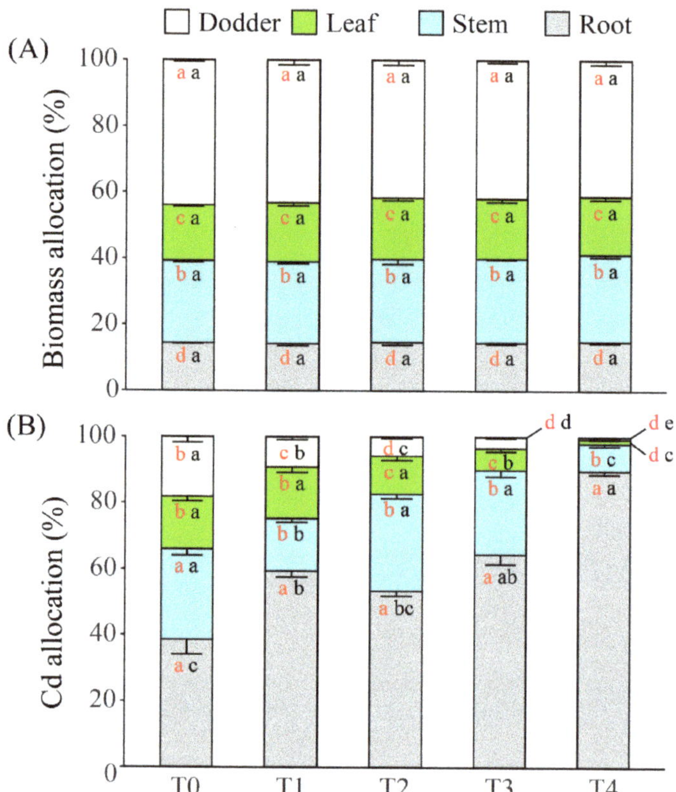

Figure 4. The allocation (i.e., proportional distribution) of (**A**) biomass and (**B**) cadmium (Cd) accumulation of various components within the soybean-dodder parasitic systems under different levels of Cd treatment. The levels of Cd treatment were applied as Hoagland solutions (50% strength) amended with 0, 1, 10, 100, and 1000 mg·L^{-1} CdCl$_2$, which are respectively marked as T0, T1, T2, T3, and T4. Different black letters within one component type indicate significant differences between Cd treatment levels of that component. Different red letters within one Cd treatment level indicate significant differences between components under that treatment level. The error bars denote 1 SE of the mean.

Cd treatment significantly changed the allocation pattern of Cd accumulation within the parasitic system (F = 149.773, $p < 0.001$). With the increase of Cd treatment levels, Cd allocation to roots continuously increased (from ca. 40% to ca. 90%), while that to leaves and dodders continuously declined (from ca. 20% to ca. 1%). The responses of Cd allocation to stems were more complex. Compared to the allocation in T0, significant reductions were only found in the lowest (T1) and highest (T4) but not the intermediate levels (T2 and T3) of Cd amendments (Figure 4B). Without Cd amendment (i.e., in T0) to the parasitic system, the rank of Cd allocation was root = stem > leaf = dodder; however, along with the intensification of Cd amendment, the rank became root > stem > leaf > dodder (Figure 4B).

3. Discussion

By conducting a controlled greenhouse experiment, we examined the transfer, accumulation as well as allocation of Cd within a soybean-dodder parasitic system. Our findings of the limited Cd allocation in dodders accompanied with the Cd transfer coefficient of the stem-dodder path always being much lower than the value of 1, clearly demonstrated no sign of Cd biomagnification through the parasitic trophic transfer from soybeans to dodders, though both the concentration and accumulation (i.e., amount) of Cd in dodders did significantly increase with the levels of Cd treatment. Below, we discuss possible reasons that may explain such interesting findings.

The absence of Cd biomagnification in dodders may be attributed to a limited transfer of Cd from soybean plants. In line with the results of numerous studies (e.g., see reviews from [9]), the majority of Cd absorbed from soil was retained in the roots of soybean plants, the process of which is believed as a primary adaptive response to reduce Cd concentration thus moderating Cd toxicity to the aboveground of plants [9]. Due to insufficient discrimination of plants between Cd ions and other essential metal ions, Cd ions can be easily taken up by root cells from soil solutions [31]. However, once Cd ions entered root cells, most of them will be complexed (e.g., chelated) by a variety of organic ligands (e.g., phytochelatins, which belong to a family of peptides rich in cysteine and are synthesized from glutathione [3,32]). Subsequently, most of these Cd compounds will be either deposited and stored in the cell walls [33] or transported and sequestered in the intracellular organelles, the vacuole in particular [3]. By doing so, the concentration of free Cd ions can be largely reduced. However, a small proportion of Cd ions together with some Cd compounds will still diffuse towards xylem via plasmodesmata, and be transported to shoots via sap flow driven by transpiration [9]. During the transportation in stem, some of the Cd ions will be further complexed by ligands and fixed in the cell walls of xylem vessels [9]. This can further reduce the availability of soluble Cd to the sinks of stems, which were leaves and dodders in our case.

No occurrence of Cd biomagnification in dodders may be further attributed to their phloem feeder characteristics, as being a holoparasite [16]. Indeed, evidence from the research of heavy metal transfer through plant-invertebrate food chains suggests that phloem suckers are less likely to biomagnify Cd than chewers during their consumption of Cd-contaminated plants, due to the limited level of mobilized Cd in phloem saps of the plants [34,35]. However, to what extent the abovementioned two explanations can hold true remains questionable. For instance, compared to invertebrate suckers which almost only rely on phloem saps, the holoparasitic dodders also take up a great amount of saps from host xylem [19], which is the main route of Cd transfer from roots to shoots within the host. Such a bi-route feeding feature could put dodders in greater risks of higher doses of Cd uptake than invertebrate phloem suckers.

Therefore, this absence of trophic enrichment of Cd in dodders may imply that the uptake of substances, at least for some heavy metals, from soybeans to dodders was a selective rather than open process. Such an explanation sounds rather conflicting to the conclusions of quite a few works which suggested that the transfer from both xylem and phloem of hosts to dodders are non-selective, since substances ranging from micromolecules (e.g., minerals and photosynthates) to macromolecules (e.g., DNA and RNA) and even to pathogens (e.g., virus and phytoplasmas) were on the list [18,19]. Anatomical analyses also confirmed that during the formation of haustorium, dodders build open connections to both xylem [36] and phloem [23] of hosts. However, throughout the literature, we do find some supports to this selective uptake hypothesis. A field study from Boyd et al. [37] observed that *C. californica* accumulated higher concentrations of potassium and phosphorus but maintained a lower concentration of nickel (Ni) than its Ni-hyperaccumulator host *Streptanthus polygaloides*. Another one from Vurro et al. [20] also showed that when parasitizing wild carrot (*Daucus carota*) in a hydroponic condition, *C. campestris* had a lower level of [Cd], while a similar concentration of copper, but a higher concentration of zinc than the host.

Nevertheless, one can still argue that these findings can be attributed to the fact that toxic heavy metals (e.g., Cd and Ni), as compared to the essential elements, in the shoots of hosts are mostly in immobilized forms that cannot be taken up by dodders. However, there is still another piece of evidence in our study that can provide further supports to the selective uptake hypothesis. That is, both dodders and soybean leaves were the xylem sinks of soybean stems, thus should compete for the same solutes (including Cd ions) in the same stem xylem transferred from the same roots. Being a super-sink [20], dodders clearly overwhelmed this competition and took away most of the solutes, as indicated by a much higher level of biomass in dodders than in leaves. Then, we would expect a higher [Cd] or at least more accumulation of Cd in dodders than in leaves. In contrast, our results clearly showed an opposite pattern, and such a pattern continuously strengthened with the intensification of Cd exposure. Thus, the trophic transfer of Cd from soybeans to dodders appeared very likely to be a selective process (unfavored or less-selected in our case) and may also partially account for the absence of Cd biomagnification in our dodders.

Of course, we should not exclude the probability that no biomagnification in dodders may be the results of experimental setups. For example, the amendments of Cd to soil were given in the middle but not the beginning of experiments, so that the period (i.e., three weeks) of Cd treatment was not long enough to generate higher levels of [Cd] in dodders than in soybean stems. In addition, the efficiency of Cd transfer to some extent also depended on soil conditions [8]. For instance, soluble Cd ions are more available for plant uptake in acidic but not alkaline soils [9]; and the extent of immobilization of soil Cd is positively correlated with the level of organic matter in soils [38]. However, there is also evidence suggesting that in the presence of chloride plants tended to take up more Cd from soil thus facilitating subsequent Cd transfer [39]. As the amendment of Cd in our experiment was given in the form of $CdCl_2$, and soybean roots had accumulated extremely high levels of Cd, the probability of our soil conditions being unsuitable for studying Cd trophic transfer is rather low.

Since Cd is extremely toxic to plants, an exposure to Cd, even at low concentrations, is expected to generate a series of detrimental effects on the growth of plants at both cellular levels (e.g., changing protein structures, reducing enzyme activities [40,41], inducing oxidative stresses [42]) and physiological levels (e.g., interrupting metabolisms [3], and interfering with water and mineral uptake [4]). Particularly for soybean, Cd exposure can significantly inhibit the photosynthetic rate by reducing chlorophyll content in leaves [43], and dampening nitrogen fixation activity by inducing nodule senescence in roots [44]. Furthermore, Cd exposure also can stimulate lignification of root cell walls, the process of which in turn can restrict the growth of roots in soybeans [45].

However, it was surprising that significant growth reductions of our soybeans only occurred in the highest level of Cd treatment. Such a 'weak' response might be since the growth of soybeans had already been strongly suppressed by dodders, the adverse effects of which largely masked the effects of Cd. Notably, to adequately test this 'mask effect' hypothesis, extra treatments of unparasitized soybeans should be included in the experimental design. The 'weak' response may also be attributed to the fact that soybeans were exposed to the Cd amendment after eight weeks of growth, by which time the plants have already passed the fast growth stage, thus yielding limited negative effects on the biomass accumulation. Indeed, most studies finding significant growth inhibitory effects had their plants treated with Cd at the seedling [43,45,46] or early growth stage [44]. Moreover, it also could be that the cultivar selected in our study happened to be a Cd-tolerant one. A growing body of literature demonstrated a genotype-dependent Cd tolerance in soybean, owning to the genotypic differences in e.g., the activities of enzymatic antioxidant system which is critical for the maintenance of membrane integrity thus redox homeostasis [47], the expressions of Cd-stress-response related MicroRNAs [48], and also the associations with arbuscular mycorrhizal fungi which play critical roles in alleviating Cd toxicity [49].

In addition to the biomass of soybean, the biomass of our dodder plants also appeared to be 'weakly' affected by Cd treatment. This could be attributed to a limited level of Cd transfer from soybean stems to dodders, so that [Cd] in dodders (except for that in T4) were still below the threshold of their body burden. In addition, similar as autotrophic plants, dodders also have evolved a series of physiological mechanisms, such as chelation and subcellular sequestration, to detoxify heavy metals [29]. For instance, the synthesis of phytochelatins plays key roles in chelating and sequestering Cd ions in plants [32]. In response to Cd exposure, dodders not only upregulate its own production of phytochelatins [20], but also take up a great amount of phytochelatins from host plants [29,50]. Such a response will strengthen their ability of Cd tolerance. Finally, as mentioned above, this 'weak' effect may also be attributed to the relatively short period of Cd exposure.

4. Materials & Methods

4.1. Plant Materials

Cuscuta chinensis, a.k.a. Chinese dodder, is an annual stem holoparasitic species characterized by rootless, leafless and string-shape yellow stems with a diameter around 1 mm. As a typical agricultural weed species, it is native to Asia and widespread in China and often parasitizes on plants of Fabaceae, Asteraceae, and Zygophyllaceae [51]. So far, the scientific community have limited their interests in the pharmaceutical values of *C. chinensis* [52,53], the biological and ecological significances of which have been overlooked until now, compared with other *Cuscuta* species, e.g., *C. australis* [27], *C. campestris* [20], *C. californica* [37], *C. japonica* [25], and *C. gronovii* [21]. A commercially available soybean (*Glycine max*) cultivar 'white in August', which is widely grown throughout China, was used as the host plants. Seeds of both Chinese dodder and soybean were obtained from local horticultural companies.

4.2. Experimental Design

This experiment was carried out in a semi-open greenhouse facility of Nanjing Forestry University from middle July (summer) to early October (autumn). Soybean seeds were surface-sterilized with a solution of 10% sodium hypochlorite for 5 min [54], and then thoroughly washed with distilled water and sown in moist sands. Three days later, germinated seedlings were transplanted into seedling trays for an initial growth of five days. Then, elder seedlings with healthy appearance and similar status were selected and transplanted to plastic pots (with a volume of 4 L) filled with commercial potting substrates (HAWITA, Germany) for experiment. The background level of Cd content in the potting substrates was around 0.133 mg·kg^{-1} (dry weight) (see the determination method in Section 4.3). To promote the growth of soybean plants, they were regularly irrigated with 100 mL Hoagland solution (50% strength) twice a week. During the whole experiment, soybean plants were carefully watered daily in a manner that soils were kept moist but without water leakage from the bottom of pots.

The infection (or parasitism) of dodders started one week after the second transplanting of soybean plants, when the light environment under soybean shoots became suitable for the germination of dodder seeds, and the subsequent host searching and haustorium induction [19] of dodder seedlings (personal experience gained from a pilot study). Specifically, dodder seeds were immersed in concentrated (98%) sulfuric acid for 15 min to promote germination (i.e., to break seed dormancy by increasing the permeability of seed coat [55]), followed by thoroughly washing the seeds with distilled water. Then, the seeds were sown to soybean pots in a manner that each pot received 20 dodder seeds which were placed on soil surface and closely surrounded the stem of the soybean plant. Once the first successful attachment (or twining) of a dodder seedling on soybean stem was observed, the rest dodder seeds or seedlings that had not yet twined on the soybean stem were removed. This can guarantee that each soybean plant was successfully parasitized by one dodder plant (personal experience gained from a pilot study). Along with the growth of dodders, their adverse impacts on soybeans continuously intensified: the growth of soybean was visually arrested; the green leaves gradually turned yellow; the flowering was stopped, and the pods were no longer produced (even if produced, they were aborted at a very early stage) (personal observation). To prevent the death of soybean plants from the parasitism of dodders before the end of the experiment, the fertilization regime was adjusted to an irrigation of 200 mL Hoagland solution (50% strength) every two days from the sixth week after the second transplanting of soybean plants.

To avoid the overly inhibitory and toxic effects from high levels of Cd treatment on soybean plants at their early growth stages (which might greatly impede the infection and early development of dodder seedlings), the amendments of Cd were started eight weeks after the second transplanting of soybean plants, when soybeans had grown strong enough to withstand both dodder parasitism and high levels of Cd stress (personal experience gained from a pilot study). Plants were exposed to five levels of Cd treatment during the every-two-day fertilization events. That is, Hoagland solutions respectively amended with 0, 1, 10, 100, and 1000 mg·L^{-1} CdCl$_2$ were given to the corresponding pots. These five levels of Cd treatment were respectively marked as T0, T1, T2, T3, and T4. In total, there were 20 pots with 20 dodder-parasitized soybean plants (i.e., four replicates per Cd treatment level). The treatment lasted for three weeks, because (i) there is evidence that significant Cd transfer from hosts to dodders can occur within two days after Cd amendment [20], and (ii) the time had just shifted from summer to autumn, gradually approaching to the end of growing season of the soybean cultivar in fields. During the three weeks, each dodder-parasitized soybean plant (i.e., soybean-dodder parasitic system) received 10 times of 200 mL CdCl$_2$–contaminated Hoagland solutions in total. Notably, based on the fact that soil Cd contamination in China was in a range between 0.003 to 9.57 mg·kg^{-1} [56] and soil bulk density in China mainly distributed around 1.4 to 1.6 kg·L^{-1} [57], our rough calculations showed that such an extent of soil Cd contamination in China can be similar to 10 times irrigation of 200 mL solution amended with 0.01 to 50 mg·L^{-1}·CdCl$_2$ into a pot filled with 4 L soil (as used in our experiment). This range thus was well included in the chosen range of Cd treatment of our experiment. The higher levels (e.g., 100 and 1000 mg·L^{-1} CdCl$_2$) of Cd treatment used here also enabled us to test whether the occurrence of Cd biomagnification in dodders is in a dose-dependent manner, e.g., biomagnification may only occur in high but not in low levels of soil Cd contamination.

4.3. Harvest and Measurements

Eleven weeks after the second transplanting of soybean plants (i.e., three weeks after the start of Cd treatment), the experiment was terminated, and the plants were harvested. Specifically, within the soybean-dodder parasitic systems, dodders were carefully separated from soybean plants. Subsequently, soybean plants were divided into roots and shoots. Roots were carefully washed free of soil, and shoots were further divided into biological stems, petioles and laminas. Regarding two reasons: (i) we found that dodders only had attached and formed haustoria into stems and petioles but not laminas of soybean plants, and (ii) both stems and petioles carried the function of resource transportation in soybeans, we pooled stems and petioles together and re-categorized them as 'stem' in the measurements and analyses. Laminas, which function as the sink receiving underground resources from stems, were also renamed as 'leaf' in the measurements and analyses. Then, all components of the soybean-dodder parasitic system (i.e., roots, stems and leaves of soybean, as well as dodders) were oven-dried at the temperature of 65 °C for three days.

The dry components were weighed, then grounded into powders and sieved through a 0.15 mm mesh for the measurements of [Cd]. Based on the test method from China National Food Safety Standard [58], [Cd] was determined with an inductively coupled plasma mass spectroscopy (iCAP RQ, Thermofisher, Waltham, MA USA) after nitric acid—hydrogen peroxide—hydrofluoric acid digestion. In addition, the background soil [Cd] in the potting substrates was previously determined. Based on China National Environmental Quality Standard for Soils [59], soil [Cd] was determined with an inductively coupled plasma atomic emission spectroscopy (iCAP 6300, Thermofisher, USA) after hydrochloric acid—nitric acid—hydrofluoric acid—perchloric acid digestion.

4.4. Statistical Analyses

Based on the biomass and [Cd] of various components, Cd accumulation ([Cd] × mass), and the allocations of biomass and Cd of the components were obtained. The transfer coefficient of Cd in different paths (i.e., root to stem, stem to leaf, and stem to dodder) were also calculated as the [Cd] ratio between sink and source components.

The effects of Cd treatment and component type on the biomass, [Cd], Cd accumulation, biomass allocation, and Cd allocation of various components within the soybean-dodder parasitic system, as well as on the Cd transfer coefficient of various paths within the system were examined using nested two-way ANOVAs with Cd treatment, component (or path) type and their interaction term as the fixed factors, and pot replicate as a random factor, followed by Tukey's post hoc tests. The effects of Cd treatment on the total mass of soybean plants were also examined using a nested one-way ANOVA with Cd treatment as the fixed factor and pot replicate as the random factor, followed by Tukey's post hoc test. In addition, the effect of receiver (i.e., sink) component type (i.e., leaf or dodder) on the correlation between stem [Cd] and its receiver [Cd] was also examined using a nested ANCOVA with stem [Cd] covariate, receiver component type and their interaction term as the fixed factors, and pot replicate as the random factor. All the statistical tests were conducted using packages 'car' [60], 'lme4' [61], 'lmerTest' [62], 'LMERConvenienceFunctions' [63], 'emmeans' [64], and 'multcomp' [65] in R v4.1.0 [66].

5. Conclusions

The current work is among the first to investigate Cd transfer from host plants to parasitic plants. We showed that among all components of the soybean-dodder parasitic system, dodders accounted for more than 40% biomass of the system but had the lowest Cd concentration and accumulated the least amount of Cd. Transfer coefficient of Cd between soybean stems and dodders was much lower than 1 and was also significantly lower than that between soybean stems and soybean leaves. These results suggested that the parasitism of stem holoparasite *C. chinensis* on Cd-contaminated hosts did not lead to Cd biomagnification. This may imply that the transfer of Cd from hosts to dodders was likely a selective process. This opinion deserves more tests since it could shed light on a new mechanism of heavy metal tolerance in parasitic plants.

Author Contributions: Conceptualization, B.J.W.C.; methodology, B.J.W.C.; validation, J.X. and X.W.; formal analysis, B.J.W.C.; investigation, J.X. and X.W.; writing—original draft preparation, B.J.W.C.; writing—review and editing, B.J.W.C.; visualization, B.J.W.C.; supervision, B.J.W.C.; project administration, B.J.W.C.; funding acquisition, B.J.W.C. All authors have read and agreed to the published version of the manuscript.

Funding: This research was funded by National Natural Science Foundation of China (32071526), and Qing Lan Project of Jiangsu Province of China.

Institutional Review Board Statement: Not applicable.

Informed Consent Statement: Not applicable.

Data Availability Statement: The data presented in this study are available in the article. Additional data are available on request from the corresponding author.

Acknowledgments: We thank Jiahe Wei, Li Huang and Zhuang Zhou for their assistance in data collection.

Conflicts of Interest: The authors declare no conflict of interest.

References

1. Hill, M.K. *Understanding Environmental Pollution*, 4th ed.; Cambridge University Press: Cambridge, UK, 2020.
2. Järup, L. Hazards of heavy metal contamination. *Br. Med. Bull.* **2003**, *68*, 167–182. [CrossRef]
3. Gratão, P.L.; Polle, A.; Lea, P.J.; Azevedo, R.A. Making the life of heavy metal-stressed plants a little easier. *Funct. Plant Ecol.* **2005**, *32*, 481–494. [CrossRef]
4. Shanmugaraj, B.M.; Malla, A.; Ramalingam, S. Cadmium stress and toxicity in plants: An overview. In *Cadmium Toxicity and Tolerance in Plants: From Physiology to Remediation*; Hasanuzzaman, M., Prasad, M.N.V., Fujita, M., Eds.; Academic Press: London, UK, 2019; pp. 1–17.

5. Wu, G.; Kang, H.; Zhang, X.; Shao, H.; Chu, L.; Ruan, C. A critical review on the bio-removal of hazardous heavy metals from contaminated soils: Issues, progress, eco-environmental concerns and opportunities. *J. Hazard. Mater.* **2010**, *174*, 1–8. [CrossRef]

6. Shahid, M.; Khalid, S.; Abbas, G.; Shahid, N.; Nadeem, M.; Sabir, M.; Aslam, M.; Dumat, C. Heavy metal stress and crop productivity. In *Crop Production and Global Environmental Issues*; Hakeem, K.R., Ed.; Springer: Cham, Switzerland, 2015; pp. 1–25.

7. Suhani, I.; Sahab, S.; Srivastava, V.; Singh, R.P. Impact of cadmium pollution on food safety and human health. *Cur. Opin. Toxicol.* **2021**, *27*, 1–7. [CrossRef]

8. McLaughlin, M.J.; Singh, B.R. Cadmium in soils and plants: A global perspective. In *Cadmium in Soils and Plants*; McLaughlin, M.J., Singh, B.R., Eds.; Springer: Dordrecht, The Netherlands, 1999; pp. 1–9.

9. Sterckeman, T.; Thomine, S. Mechanisms of cadmium accumulation in plants. *Crit. Rev. Plant Sci.* **2020**, *39*, 322–359. [CrossRef]

10. Abedi, T.; Mojiri, A. Cadmium uptake by wheat (*Triticum aestivum* L.): An overview. *Plants* **2020**, *9*, 500. [CrossRef]

11. Waalkes, M.P. Cadmium carcinogenesis in review. *J. Inorg. Biochem.* **2000**, *79*, 241–244. [CrossRef]

12. Hunter, B.A.; Johnson, M.S. Food chain relationships of copper and cadmium in contaminated grassland ecosystems. *Oikos* **1982**, *38*, 108–117. [CrossRef]

13. Li, T.; Chang, Q.; Yuan, X.; Li, J.; Ayoko, G.A.; Frost, R.L.; Chen, H.; Zhang, X.; Song, Y.; Song, W. Cadmium transfer from contaminated soils to the human body through rice consumption in southern Jiangsu Province, China. *Environ. Sci. Process. Impacts* **2017**, *19*, 843–850. [CrossRef]

14. Smith, T.M.; Smith, R.L. *Elements of Ecology: International Edition*, 6th ed.; Pearson Education, Benjamin Cummings: San Francisco, CA, USA, 2006.

15. Pennings, S.C.; Callaway, R.M. Parasitic plants: Parallels and contrasts with herbivores. *Oecologia* **2002**, *131*, 479–489. [CrossRef]

16. Twyford, A.D. Parasitic plants. *Cur. Biol.* **2018**, *28*, R857–R859. [CrossRef] [PubMed]

17. Parker, C. Parasitic weeds: A world challenge. *Weed Sci.* **2012**, *60*, 269–276. [CrossRef]

18. Kim, G.; Westwood, J.H. Macromolecule exchange in *Cuscuta*–host plant interactions. *Curr. Opin. Plant Biol.* **2015**, *26*, 20–25. [CrossRef]

19. Furuhashi, T.; Furuhashi, K.; Weckwerth, W. The parasitic mechanism of the holostemparasitic plant *Cuscuta*. *J. Plant Interact.* **2011**, *6*, 207–219. [CrossRef]

20. Vurro, E.; Ruotolo, R.; Ottonello, S.; Elviri, L.; Maffini, M.; Falasca, G.; Zanella, L.; Altamura, M.M.; Sanità di Toppi, L. Phytochelatins govern zinc/copper homeostasis and cadmium detoxification in *Cuscuta campestris* parasitizing *Daucus carota*. *Environ. Exp. Bot.* **2011**, *72*, 26–33. [CrossRef]

21. Touchette, B.W.; Feely, S.; McCabe, S. Elevated nutrient content in host plants parasitized by swamp dodder (*Cuscuta gronovii*): Evidence of selective foraging by a holoparasitic plant? *Plant Biosyst.* **2021**. [CrossRef]

22. Vaughn, K.C. Conversion of the searching hyphae of dodder into xylic and phloic hyphae: A cytochemical and immunocytochemical investigation. *Int. J. Plant Sci.* **2006**, *167*, 1099–1114. [CrossRef]

23. Birschwilks, M.; Haupt, S.; Hofius, D.; Neumann, S. Transfer of phloem-mobile substances from the host plants to the holoparasite *Cuscuta* sp. *J. Exp. Bot.* **2006**, *57*, 911–921. [CrossRef] [PubMed]

24. Shen, H.; Hong, L.; Ye, W.; Cao, H.; Wang, Z. The influence of the holoparasitic plant *Cuscuta campestris* on the growth and photosynthesis of its host *Mikania micrantha*. *J. Exp. Bot.* **2007**, *58*, 2929–2937. [CrossRef]

25. Furuhashi, T.; Kojima, M.; Sakakibara, H.; Fukushima, A.; Hirai, M.Y.; Furuhashi, K. Morphological and plant hormonal changes during parasitization by *Cuscuta japonica* on *Momordica charantia*. *J. Plant Interact.* **2014**, *9*, 220–232. [CrossRef]

26. Zhuang, H.; Li, J.; Song, J.; Hettenhausen, C.; Schuman, M.C.; Sun, G.; Zhang, C.; Li, J.; Song, D.; Wu, J. Aphid (*Myzus persicae*) feeding on the parasitic plant dodder (*Cuscuta australis*) activates defense responses in both the parasite and soybean host. *New Phytol.* **2018**, *218*, 1586–1596. [CrossRef]

27. Liu, N.; Shen, G.; Xu, Y.; Liu, H.; Zhang, J.; Li, S.; Li, J.; Zhang, C.; Qi, J.; Wang, L.; et al. Extensive inter-plant protein transfer between *Cuscuta* parasites and their host plants. *Mol. Plant* **2020**, *13*, 573–585. [CrossRef]

28. Zagorchev, L.; Stöggl, W.; Teofanova, D.; Li, J.; Kranner, I. Plant parasites under pressure: Effects of abiotic stress on the interactions between parasitic plants and their hosts. *Int. J. Mol. Sci.* **2021**, *22*, 7418. [CrossRef]

29. Srivastava, S.; Tripathi, R.D.; Dwivedi, U.N. Synthesis of phytochelatins and modulation of antioxidants in response to cadmium stress in *Cuscuta reflexa*—An angiospermic parasite. *J. Plant Physiol.* **2004**, *161*, 665–674. [CrossRef]

30. Naikoo, M.I.; Raghib, F.; Dar, M.I.; Khan, F.A.; Hessini, K.; Ahmad, P. Uptake, accumulation and elimination of cadmium in a soil—Faba bean (*Vicia faba*)—Aphid (*Aphis fabae*)—Ladybird (*Coccinella transversalis*) food chain. *Chemosphere* **2021**, *279*, 130522. [CrossRef]

31. Cataldo, D.A.; Garland, T.R.; Wildung, R.E. Cadmium distribution and chemical fate in soybean plants. *Plant Physiol.* **1981**, *68*, 835–839. [CrossRef]

32. Yadav, S.K. Heavy metals toxicity in plants: An overview on the role of glutathione and phytochelatins in heavy metal stress tolerance of plants. *S. Afr. J. Bot.* **2010**, *76*, 167–179. [CrossRef]

33. Vázquez, S.; Goldsbrough, P.; Carpena, R.O. Assessing the relative contributions of phytochelatins and the cell wall to cadmium resistance in white lupin. *Physiol. Plant.* **2006**, *128*, 487–495. [CrossRef]

34. Konopka, J.K.; Hanyu, K.; Macfie, S.M.; McNeil, J.N. Does the response of insect herbivores to cadmium depend on their feeding strategy? *J. Chem. Ecol.* **2013**, *39*, 546–554. [CrossRef]

35. Butt, A.; Quratul, A.; Rehman, K.; Khan, M.X.; Hesselberg, T. Bioaccumulation of cadmium, lead, and zinc in agriculture-based insect food chains. *Environ. Monit. Assess.* **2018**, *190*, 698. [CrossRef]

36. Dawson, J.H.; Musselman, L.J.; Wolswinkel, P.; Dörr, I. Biology and control of *Cuscuta*. *Rev. Weed Sci.* **1994**, *6*, 265–317.

37. Boyd, R.S.; Martens, S.N.; Davis, M.A. The nickel hyperaccumulator *Streptanthus polygaloides* (Brassicaceae) is attacked by the parasitic plant *Cuscuta californica* (Cuscutaceae). *Madroño* **1999**, *46*, 92–99.

38. Lin, Y.-F.; Hassan, Z.; Talukdar, S.; Schat, H.; Aarts, M.G.M. Expression of the ZNT1 zinc transporter from the metal hyperaccumulator *Noccaea caerulescens* confers enhanced zinc and cadmium tolerance and accumulation to *Arabidopsis thaliana*. *PLoS ONE* **2016**, *11*, e0149750. [CrossRef]

39. Weggler, K.; McLaughlin, M.J.; Graham, R.D. Effect of chloride in soil solution on the plant availability of biosolid-borne cadmium. *J. Environ. Qual.* **2004**, *33*, 496–504. [CrossRef]

40. He, L.; Wang, X.; Feng, R.; He, Q.; Wang, S.; Liang, C.; Yan, L.; Bi, Y. Alternative pathway is involved in nitric oxide-enhanced tolerance to cadmium stress in barley roots. *Plants* **2019**, *8*, 557. [CrossRef]

41. Mendoza-Cózatl, D.; Loza-Tavera, H.; Hernández-Navarro, A.; Moreno-Sánchez, R. Sulfur assimilation and glutathione metabolism under cadmium stress in yeast, protists and plants. *FEMS Microbiol. Rev.* **2005**, *29*, 653–671. [CrossRef]

42. Schutzendubel, A.; Schwanz, P.; Teichmann, T.; Gross, K.; Langenfeld-Heyser, R.; Godbold, D.L.; Polle, A. Cadmium-induced changes in antioxidative systems, hydrogen peroxide content, and differentiation in scots pine roots. *Plant Physiol.* **2001**, *127*, 887–898. [CrossRef]

43. Xue, Z.-C.; Gao, H.-Y.; Zhang, L.-T. Effects of cadmium on growth, photosynthetic rate and chlorophyll content in leaves of soybean seedlings. *Biol. Plant.* **2013**, *57*, 587–590. [CrossRef]

44. Balestrasse, K.B.; Gallego, S.M.; Tomaro, M.L. Cadmium-induced senescence in nodules of soybean (*Glycine max* L.) plants. *Plant Soil* **2004**, *262*, 373–381. [CrossRef]

45. Finger-Teixeira, A.; Lucio Ferrarese, M.d.L.; Ricardo Soares, A.; da Silva, D.; Ferrarese-Filho, O. Cadmium-induced lignification restricts soybean root growth. *Ecotoxicol. Environ. Saf.* **2010**, *73*, 1959–1964. [CrossRef]

46. Chen, Y.X.; He, Y.F.; Yang, Y.; Yu, Y.L.; Zheng, S.J.; Tian, G.M.; Luo, Y.M.; Wong, M.H. Effect of cadmium on nodulation and N_2-fixation of soybean in contaminated soils. *Chemosphere* **2003**, *50*, 781–787. [CrossRef]

47. Alyemeni, M.N.; A., A.M.; Wijaya, L.; Alam, P.; Ahmad, P. Contrasting tolerance among soybean genotypes subjected to different levels of cadmium stress. *Pak. J. Bot.* **2017**, *49*, 903–911.

48. Fang, X.; Zhao, Y.; Ma, Q.; Huang, Y.; Wang, P.; Zhang, J.; Nian, H.; Yang, C. Identification and comparative analysis of cadmium tolerance-associated miRNAs and their targets in two soybean genotypes. *PLoS ONE* **2013**, *8*, e81471. [CrossRef]

49. Cui, G.; Ai, S.; Chen, K.; Wang, X. Arbuscular mycorrhiza augments cadmium tolerance in soybean by altering accumulation and partitioning of nutrient elements, and related gene expression. *Ecotoxicol. Environ. Saf.* **2019**, *171*, 231–239. [CrossRef]

50. Zagorchev, L.; Albanova, I.; Tosheva, A.; Li, J.; Teofanova, D. Salinity effect on *Cuscuta campestris* Yunck. parasitism on *Arabidopsis thaliana* L. *Plant Physiol. Biochem.* **2018**, *132*, 408–414. [CrossRef]

51. Fang, R.-C.; Staples, G. Convolvulaceae. In *Flora of China, Volume 16: Gentianaceae through Boraginaceae*; Wu, C.Y., Raven, P.H., Eds.; Missouri Botanical Garden Press: St. Louis, MI, USA, 1995; pp. 271–325.

52. Bao, X.; Wang, Z.; Fang, J.; Li, X. Structural features of an immunostimulating and antioxidant acidic polysaccharide from the seeds of *Cuscuta chinensis*. *Planta Med.* **2002**, *68*, 237–243. [CrossRef]

53. Donnapee, S.; Li, J.; Yang, X.; Ge, A.-H.; Donkor, P.O.; Gao, X.-M.; Chang, Y.-X. *Cuscuta chinensis* Lam.: A systematic review on ethnopharmacology, phytochemistry and pharmacology of an important traditional herbal medicine. *J. Ethnopharmacol.* **2014**, *157*, 292–308. [CrossRef]

54. Chen, B.J.W.; Huang, L.; During, H.J.; Wang, X.; Wei, J.; Anten, N.P.R. No neighbour-induced increase in root growth of soybean and sunflower in mesh-divider experiments after controlling for nutrient concentration and soil volume. *AoB Plants* **2021**, *13*, plab020. [CrossRef]

55. Hutchison, J.M.; Ashton, F.M. Effect of desiccation and scarification on the permeability and structure of the seed coat of *Cuscuta campestris*. *Am. J. Bot.* **1979**, *66*, 40–46. [CrossRef]

56. Wang, L.; Cui, X.; Cheng, H.; Chen, F.; Wang, J.; Zhao, X.; Lin, C.; Pu, X. A review of soil cadmium contamination in China including a health risk assessment. *Environ. Sci. Pollut. Res.* **2015**, *22*, 16441–16452. [CrossRef]

57. Han, G.-Z.; Zhang, G.-L.; Gong, Z.-T.; Wang, G.-F. Pedotransfer functions for estimating soil bulk density in China. *Soil Sci.* **2012**, *177*, 158–164. [CrossRef]

58. National Health Commission of the PRC; National Medical Products Administration of the PRC. China National Food Safety Standard. In *Determination of Elements in Food*; GB5009.268-2016; 2016; Available online: https://sppt.cfsa.net.cn:8086/staticPages/DBE4CA28-C983-42EC-AE60-5EB9A0ED008A.html?clicks=384 (accessed on 6 October 2020).

59. State Environmental Protection Administration of the PRC; State Bureau of Quality and Technical Supervision of the PRC. Environmental Quality Standard for Soils. 1995. GB15618-1995. Available online: https://www.chinesestandard.net/Related.aspx/GB15618-1995 (accessed on 6 October 2020).

60. Fox, J.; Weisberg, S. *An R Companion to Applied Regression*, 3rd ed.; Sage Publishing: Thousand Oaks, CA, USA, 2019.

61. Bates, D.; Mächler, M.; Bolker, B.; Walker, S. Fitting linear mixed-effects models using lme4. *J. Stat. Softw.* **2015**, *67*, 1–48. [CrossRef]

62. Kuznetsova, A.; Brockhoff, P.B.; Christensen, R.H.B. lmerTest package: Tests in linear mixed effects models. *J. Stat. Softw.* **2017**, *82*, 1–26. [CrossRef]

63. Tremblay, A.; Ransijn, J. LMERConvenienceFunctions: Model Selection and Post-Hoc Analysis for (G)LMER Models. R package version 3.0. Available online: https://CRAN.R-project.org/package=LMERConvenienceFunctions (accessed on 6 October 2020).
64. Lenth, R.V. Emmeans: Estimated Marginal Means, Aka Least-Squares Means. R Package Version 1.6.3. Available online: https://CRAN.R-project.org/package=emmeans (accessed on 29 September 2021).
65. Hothorn, T.; Bretz, F.; Westfall, P. Simultaneous inference in general parametric models. *Biom. J.* **2008**, *50*, 346–363. [CrossRef]
66. R Core Team. *R: A Language and Environment for Statistical Computing*; R Foundation for Statistical Computing: Vienna, Austria, 2021.

Article

Mobile Host mRNAs Are Translated to Protein in the Associated Parasitic Plant *Cuscuta campestris*

So-Yon Park [1,†,‡], Kohki Shimizu [2,†], Jocelyn Brown [1], Koh Aoki [2,*] and James H. Westwood [1,*]

1   School of Plant and Environmental Sciences, Virginia Tech, Blacksburg, VA 24061, USA;
    parksoy@missouri.edu (S.-Y.P.); jocebrown21@gmail.com (J.B.)
2   Graduate School of Life and Environmental Sciences, Osaka Prefecture University, 1-1 Gakuen-cho, Naka-ku,
    Sakai 599-8531, Japan; kohki.72h@gmail.com
*   Correspondence: kaoki@plant.osakafu-u.ac.jp (K.A.); westwood@vt.edu (J.H.W.)
†   Co-first author. These authors contributed equally to the work.
‡   Present addresses: Division of Plant Science, University of Missouri, Columbia, MO 65211, USA.

Abstract: *Cuscuta* spp. are obligate parasites that connect to host vascular tissue using a haustorium. In addition to water, nutrients, and metabolites, a large number of mRNAs are bidirectionally exchanged between *Cuscuta* spp. and their hosts. This trans-specific movement of mRNAs raises questions about whether these molecules function in the recipient species. To address the possibility that mobile mRNAs are ultimately translated, we built upon recent studies that demonstrate a role for transfer RNA (tRNA)-like structures (TLSs) in enhancing mRNA systemic movement. *C. campestris* was grown on *Arabidopsis* that expressed a β-glucuronidase (*GUS*) reporter transgene either alone or in *GUS-tRNA* fusions. Histochemical staining revealed localization in tissue of *C. campestris* grown on *Arabidopsis* with *GUS-tRNA* fusions, but not in *C. campestris* grown on *Arabidopsis* with *GUS* alone. This corresponded with detection of *GUS* transcripts in *Cuscuta* on *Arabidopsis* with *GUS-tRNA*, but not in *C. campestris* on *Arabidopsis* with *GUS* alone. Similar results were obtained with *Arabidopsis* host plants expressing the same constructs containing an endoplasmic reticulum localization signal. In *C. campestris*, GUS activity was localized in the companion cells or phloem parenchyma cells adjacent to sieve tubes. We conclude that host-derived *GUS* mRNAs are translated in *C. campestris* and that the TLS fusion enhances RNA mobility in the host-parasite interactions.

Keywords: parasitic plants; *Cuscuta*; tRNA; mobile mRNA

Citation: Park, S.-Y.; Shimizu, K.; Brown, J.; Aoki, K.; Westwood, J.H. Mobile Host mRNAs Are Translated to Protein in the Associated Parasitic Plant *Cuscuta campestris*. *Plants* **2022**, *11*, 93. https://doi.org/10.3390/plants11010093

Academic Editors: Evgenia Dor and Yaakov Goldwasser

Received: 17 November 2021
Accepted: 23 December 2021
Published: 28 December 2021

Publisher's Note: MDPI stays neutral with regard to jurisdictional claims in published maps and institutional affiliations.

1. Introduction

Cuscuta spp. (dodders) are holoparasitic plants that attack a broad range of hosts, and are capable of causing substantial agricultural losses [1]. *Cuscuta* plants typically consist of yellow or orange stems, lacking roots or developed leaves. They connect by coiling around host stems, petioles, and leaves, and at these points of contact they develop haustoria, which are unique structures that grow invasively into the host to form a continuum with the host's xylem and phloem tissues [2]. The haustorium functions to feed the parasite though uptake of water, sugars, and other nutrients, but is also capable of facilitating exchange of macromolecules including proteins [3], mRNAs [4], microRNAs [5], and possibly even DNAs, as implicated by horizontal gene transfer [6]. Movement of each of these classes of macromolecules raises many questions regarding the exchange of signals between host and parasite, but the least understood are arguably mRNAs, for which little is known about their mechanisms of movement, fate, and function in the plant-plant interaction. In particular, it is important to understand whether mobile mRNAs from the host are able to be translated into protein after arriving in the parasite, as this would provide a powerful mechanism for transmission of proteins that otherwise would be unable to move between the organisms.

Plants have evolved the ability to transport RNAs over long distances in the phloem. These non-cell-autonomous mRNAs are thought to function in coordinating plant development and response to stress [7]. Several mobile mRNAs have been demonstrated to affect the phenotype of the destination tissue, including *Flowering locus T (FT)*, for which mobile protein and mRNA move from leaf phloem into shoot apical meristem to promote flowering [8]. Other well-characterized long-distance mobile mRNAs associated with phenotypes are a fusion of *pyrophosphate-dependent phosphofructokinase* with *LeT6* in tomato [9], the *BEL5* transcription factor from potato [10], and *Gibberellic-Acid insensitive* [11], among others [12,13]. Recent studies have identified large numbers of mobile cellular mRNAs through hetero-grafting combined with high-throughput sequencing technologies [14–16]. The large-scale exchange of mRNAs between *Cuscuta* plants and their hosts suggests that they are able to tap into this system, although the biological significance is not yet clear [17].

Another unsolved mystery is the mechanism(s) by which the cell-to-cell movement of mRNAs is regulated in plants. Studies have indicated multiple factors that contribute to mRNA mobility, including sequences of the $3'$ and $5'$ untranslated regions (UTRs) [18], and the presence of methylated cytosine bases in the mRNA coding sequence, or UTRs [19]. Furthermore, transfer RNA (tRNA) sequences or tRNA-like structures (TLSs) in the $3'$ UTR of an mRNA were found to increase systemic mobility of associated mRNAs in plants [13,20]. In the latter work, Zhang et al. [20] added tRNA sequences to the β-glucuronidase (GUS) protein coding sequence and showed that they were sufficient to promote GUS mRNA mobility across *Arabidopsis* graft junctions. GUS enzyme activity was detected in the recipient tissue; in this case, wild type roots grafted to shoots expressing the *35S:GUS-tRNA* transgene. They also demonstrated that the mobile *GUS-tRNA* mRNA was translated to protein in the roots. Not all tRNAs conferred mobility to associated mRNAs, so there is specificity in the system. For example, the tRNAs for methionine (tRNA^{Met}) and glycine (tRNA^{Gly}) conferred mobility, while the isoleucine tRNA (tRNA^{Ile}) did not. The three-dimensional structure of the TLSs was shown to be important, as indicated by the finding that certain mutations of the hairpin loop structures affect mobility, as deletion of A and T loops of tRNA^{Met} (tRNA$^{Met-dAT}$) abolished movement, while deletion of D and T loops (tRNA$^{Met-dDT}$) retained mobility.

Our long-term objective is to understand the mechanisms by which *Cuscuta* spp. interact with their hosts, and specifically the role of RNAs in the interaction. Recent work by Liu et al. [3] suggested that protein movement between hosts and *C. australis* takes place primarily by direct protein movement, without need for an mRNA intermediary. In this paper, we address two central questions: (1) Does a tRNA fusion system that confers cell-cell mobility on GUS gene mRNAs in *Arabidopsis* also enable it to traffic into *C. campestris*? (2) Is such a mobile GUS mRNA translated into protein in *C. campestris*? Indeed, we have found that tRNA fused to the GUS gene facilitates the movement of GUS mRNA and results in GUS enzyme activity in *C. campestris* haustoria, stems, floral organs, phloem, and apical termini of sieve tubes. These results support the idea that the transported *GUS-tRNA* mRNA from *Arabidopsis* host plants is translated in *C. campestris* cells.

2. Results

2.1. tRNA Fusions Influence Mobility of GUS Activity

We used transgenic *Arabidopsis* plants expressing *GUS* either with or without *tRNA* fusions and assayed the movement of GUS activity into attached *C. campestris*. For this experiment, *C. campestris* stems parasitizing *Arabidopsis* floral shoots were sectioned as a unit and stained to reveal GUS activity (Figure 1A). As a negative control, we examined *C. campestris* growing on nontransgenic *Arabidopsis* because the related species *C. pentagona* has been reported to have endogenous GUS activity [21]. Unlike *C. pentagona*, no GUS activity was detected in wild type *C. campestris* (Figure 1B). *C. campestris* was then grown on *Arabidopsis* with *35S:GUS* or *35S:GUS-tRNA*^{Met} transgenes and again sectioned and stained to reveal GUS activity. No GUS activity was detected in *C. campestris* expressing GUS without the tRNA sequence (Figure 1C,D) but was evident in *C. campestris* parasitizing hosts expressing *GUS-tRNA^{Met}* (Figure 1E,F). These results indicate that the presence of tRNA motif promotes mobility of GUS activity from host to *C. campestris*, similar to its function in *Arabidopsis* grafting experiments [20].

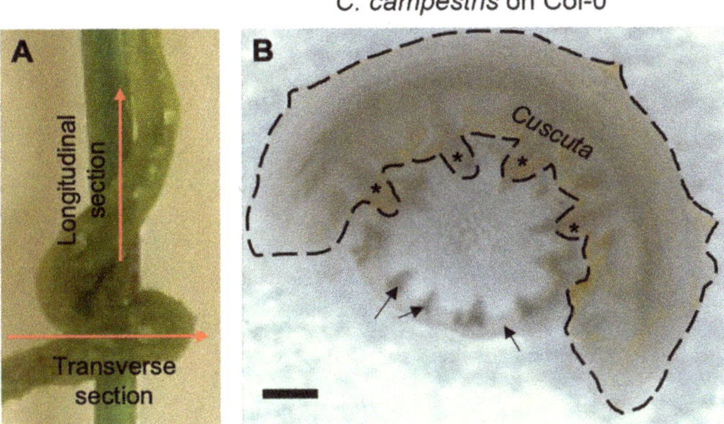

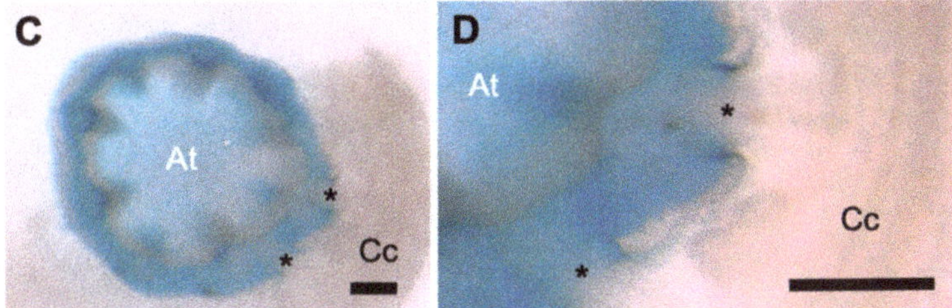

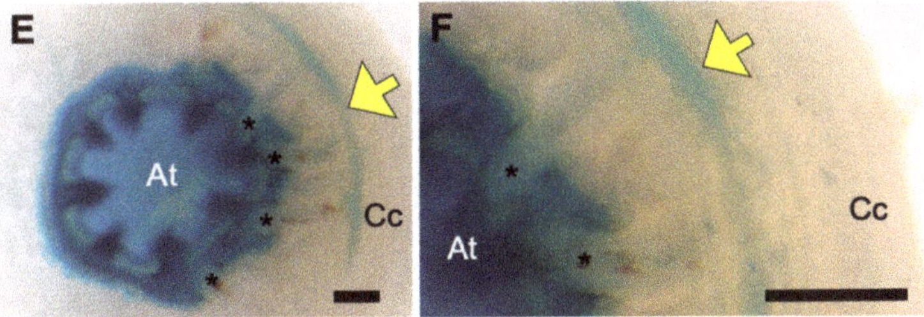

Figure 1. Histochemical localization of β-glucuronidase. Haustoria between *Arabidopsis* and *Cuscuta campestris* were transversely cross-sectioned (as indicated by the red arrow) (**A**). *C. campestris* was inoculated on stems of 3-week-old *Arabidopsis* plants; wild type (WT) (**B**), *35S:GUS* (**C,D**), and *35S:GUS-tRNA^{Met}* (**E,F**). D and F are high-magnification images of C and E, respectively. (**B–F**). The blue color of GUS activity in *C. campestris* is indicated by yellow arrows (**E,F**). Asterisks indicate haustoria. Scale bar: 500 μm.

Considering the open exchange of materials between *Cuscuta* spp. and their hosts, it is important to use extra caution in judging whether GUS moves as a protein, as opposed to an mRNA that is subsequently translated into protein. Although GUS has been considered to be a non-mobile protein, having been used for decades as a cell- and tissue-specific indicator of gene expression [22], it has been proposed to be mobile from host plants to *C. australis* [3]. Therefore, to further restrict GUS protein mobility, we fused a sequence encoding the endoplasmic reticulum (ER) signal peptide to the *GUS* gene construct. Previous studies have shown that ER targeting peptides are sufficient to block GFP protein movement [23,24]. Additionally, we used tRNA variants that were shown to differ in ability to confer mobility on mRNAs in grafted *Arabidopsis* [20]. Thus, in addition to using the non-ER localized GUS constructs, we generated transgenic *Arabidopsis* expressing *35S:ER-GUS*, *35S:ER-GUS-tRNA^{Met}* and *35S:ER-GUS-tRNA^{Met-dDT}*, as well as others derived from the constructs reported by Zhang et al. (2016). Transgenic *Arabidopsis* plants were confirmed to show strong GUS activity using the fluorescent 4-MUG assay, while wild type plants had negligible activity (Supplemental Table S1).

C. campestris was grown on the *Arabidopsis* plants expressing GUS with or without tRNA fusions and with or without ER localization signals. The parasite stem was removed from the host and the haustoria regions were sectioned longitudinally and transversely before staining to detect GUS activity. No GUS activity was detected in *C. campestris* parasitizing hosts with *35S:GUS* or *35S:ER-GUS* (Figure 2A,B,G,H). However, the blue dye indicative of GUS activity was evident in *C. campestris* parasitizing hosts with tRNA fusions to the *GUS* gene: *35S:GUS-tRNA^{Met}*, *35S:GUS-tRNA^{Met-dDT}*, *35S:ER-GUS-tRNA^{Met}*, and *35S:ER-GUS-tRNA^{Met-dDT}* (Figure 2C–F,I–L). This pattern was confirmed by counting the number of haustoria showing GUS activity on these and additional transgenic lines. Haustoria from negative controls (wild type Col-0, *35S:empty*, *35S:GUS*, and *35S:ER-GUS*) never showed GUS enzyme activity (Table 1). In contrast, 30% to 80% of *C. campestris* haustorial regions parasitizing *Arabidopsis* GUS lines with *tRNA^{Met}*, *tRNA^{Met-dDT}*, *tRNA^{Gly}*, and *tRNA^{Ile}* fusions showed GUS activity. Furthermore, 30% to 39% of *C. campestris* haustoria growing on hosts with ER-GUS-tRNAs showed GUS enzyme activity. The one exception was a lack of GUS enzyme activity in *C. campestris* growing on *Arabidopsis* expressing *35S:GUS-tRNA^{Met-dAT}*, although this is consistent with a lack of mobility reported for this construct in the *Arabidopsis* grafting assay [20].

Table 1. Percent of *Cuscuta campestris* haustoria showing GUS enzyme activity.

Arabidopsis Lines	Number of *Cuscuta* Haustoria		Total Number of Samples	% with GUS Detection
	GUS Detected	**No GUS**		
Wild type Col-0	**0**	13	13	0
35S:empty (pEarleygate100)	0	10	10	0
35S:GUS	0	35	35	0
35S:GUS-tRNA^{MET}	39	10	49	80
35S:GUS-tRNA^{MET dDT}	14	20	34	41
35S:GUS-tRNA^{Gly}	12	20	32	38
35S:GUS-tRNA^{Ile}	11	26	37	30
35S:GUS-tRNA^{MET dAT}	0	12	12	0
35S:ER-GUS	0	12	12	0
35S:ER-GUS-tRNA^{MET}	9	17	26	35
35S:ER-GUS-tRNA^{MET dDT}	13	20	33	39
35S:ER-GUS-tRNA^{Gly}	11	17	28	39
35S:ER-GUS-tRNA^{Ile}	8	19	27	30

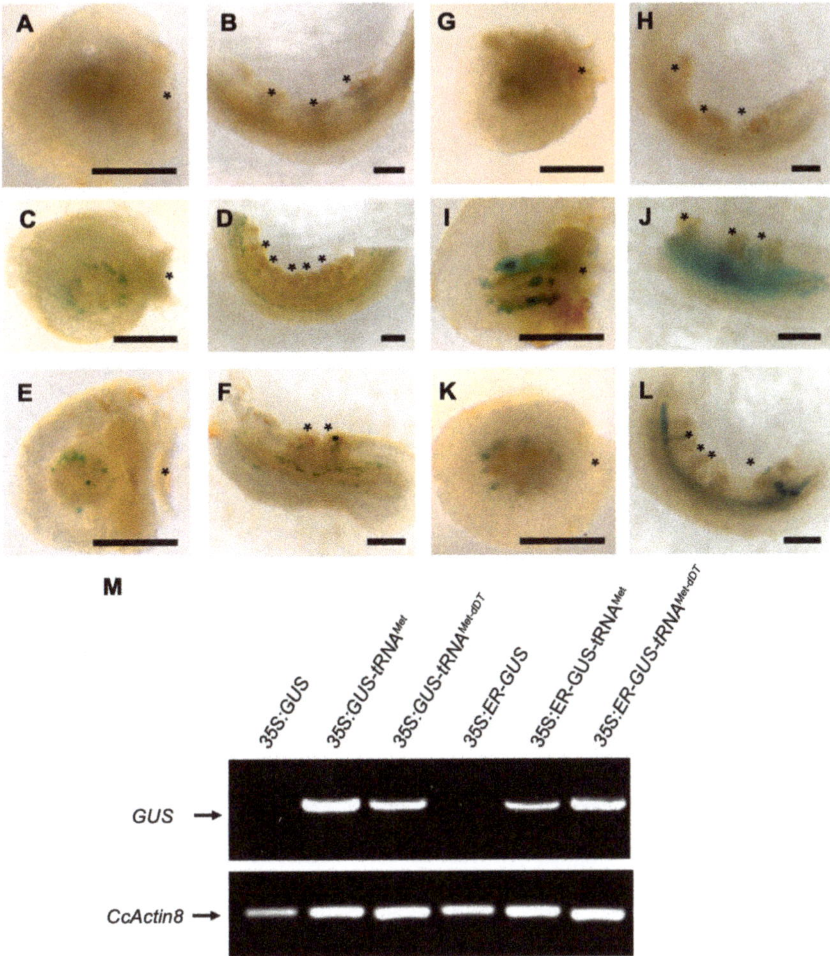

Figure 2. GUS mRNA movement. *Cuscuta campestris* was inoculated on stems of three-week-old *Arabidopsis*; *35:GUS* (**A**,**B**), *35S:GUS-tRNA^{Met}* (**C**,**D**), *35S:GUS-tRNA$^{Met-dDT}$* (**E**,**F**), *35:ER-GUS* (**G**,**H**), *35S:ER-GUS-tRNA^{Met}* (**I**,**J**), and *35S:ER-GUS-tRNA$^{Met-dDT}$* (**K**,**L**). Haustoria between *Arabidopsis* and *C. campestris* were longitudinally (**A**,**C**,**E**,**G**,**I**,**K**) and transversely (**B**,**D**,**F**,**H**,**J**,**L**) cross-sectioned. Asterisks indicate haustoria. GUS mRNA was detected in *C. campestris* stems on the *35S:GUS-tRNA^{-Met}*, *35S:GUS-tRNA$^{-Met-dDT}$*, *35S:ER-GUS-tRNA^{-Met}*, and *35S:ER-GUS-tRNA$^{-Met-dDT}$*. *C. campestris Actin8* (*CcActin8*) was used as a reference gene (**M**). Scale bar: 500 μm.

2.2. GUS mRNA in C. campestris Is Associated with GUS-tRNA Fusions

We investigated the mobility of *GUS* mRNA from *Arabidopsis* plants expressing *GUS* with or without tRNA sequences. To avoid any possibility of contamination from parasite tissues in close contact with the host, total RNA was extracted from *C. campestris* stem more than 1 cm away from the haustoria. RT-PCR was used to detect mRNAs from the *GUS* gene constructs and *C. campestris* actin gene (*CcActin8*) as a positive control. While *CcActin8* was amplified from all samples, *GUS* mRNAs were only amplified from parasite tissues where *Cuscuta* was growing on *Arabidopsis* expressing tRNA fusions: *GUS-tRNA^{Met}*, *GUS-tRNA$^{Met\text{-}dDT}$*, *ER-GUS-tRNA^{Met}*, and *ER-GUS-tRNA$^{Met\text{-}dDT}$* (Figure 2M).

2.3. GUS mRNA Moves Long Distances in C. campestris and GUS Activity Is Localized in Phloem Cells

To investigate the distribution of GUS protein in *C. campestris*, stems of the parasite were sectioned at increasing distances from the haustorial region (Supplemental Figure S1A). GUS enzyme activity was strongly expressed in the *Arabidopsis 35S:GUS-tRNA^{Met}* host stems (Supplemental Figure S1B,C,E,F). GUS activities were detected in *C. campestris* stems near the haustoria regions (Supplemental Figure S1C,F), as well as from 0.7 cm to 12 cm away (Supplemental Figure S1D,G–I). Quantitative RT-PCR Analyses of mRNAs from the same experiment indicated the presence of mobile *GUS-tRNA^{Met}* and *GUS-tRNA-$^{Met\text{-}dDT}$* transcripts from the entire length of the *C. campestris* stem (Supplemental Figure S1K,L).

To further localize the presence of the GUS enzyme, we assayed flowers of *C. campestris* grown on *35S:GUS-tRNA^{Met}*. GUS activity was detected at the base of floral buds located 4 to 6 cm away from the haustoria (Figure 3A). Specifically, GUS was observed in the peduncle and the base of, but not inside, the *C. campestris* ovary (Figure 3B,C). GUS activity was also detected in the vascular tissues at the base of the apical tip of *C. campestris* grown on *35S:GUS-tRNA^{Met}* and *35S:GUS-tRNA$^{Met\text{-}dDT}$* expressing *Arabidopsis* (Figure 3E,F). As in the flower, GUS activity was not detected in the meristematic region. Longitudinal (Figure 3H,I) and transverse (Figure 3J,K) sections showed that GUS activity was not co-localized with xylem. In further support of this observation, sequential staining for GUS activity, followed by phloroglucinol-HCl staining of lignin in xylem cells [25], indicated that for *C. campestris* growing on *Arabidopsis 35S:GUS-tRNA^{Met}* plants the GUS signals were detected more centrally in the *C. campestris* stem than the lignin staining (Supplemental Figure S2).

To test whether GUS activity was localized in the phloem, we performed double staining for GUS activity and callose deposition that is indicative of sieve plates. GUS signals were detected first in *C. campestris* grown on *Arabidopsis 35S:GUS-tRNA^{Met}* (Figure 4A); then the same sections were transferred to a confocal microscopy to identify GUS-stained cells by transmission image (Figure 4B) and stained with aniline blue to visualize callose deposition on the sieve plates of sieve tubes (Figure 4C). GUS activity was localized in the array of cells next to sieve tubes containing aniline blue-stained sieve plates (Figure 4D). Essentially, the same localization patterns of GUS activity and sieve tubes were obtained in *C. campestris* grown on *Arabidopsis 35S:GUS-tRNA$^{Met\text{-}dDT}$* and *35S:GUS-tRNA^{Gly}* (Supplemental Figure S3). These results suggest that GUS proteins were localized in the companion cells or phloem parenchyma cells adjacent to sieve tubes.

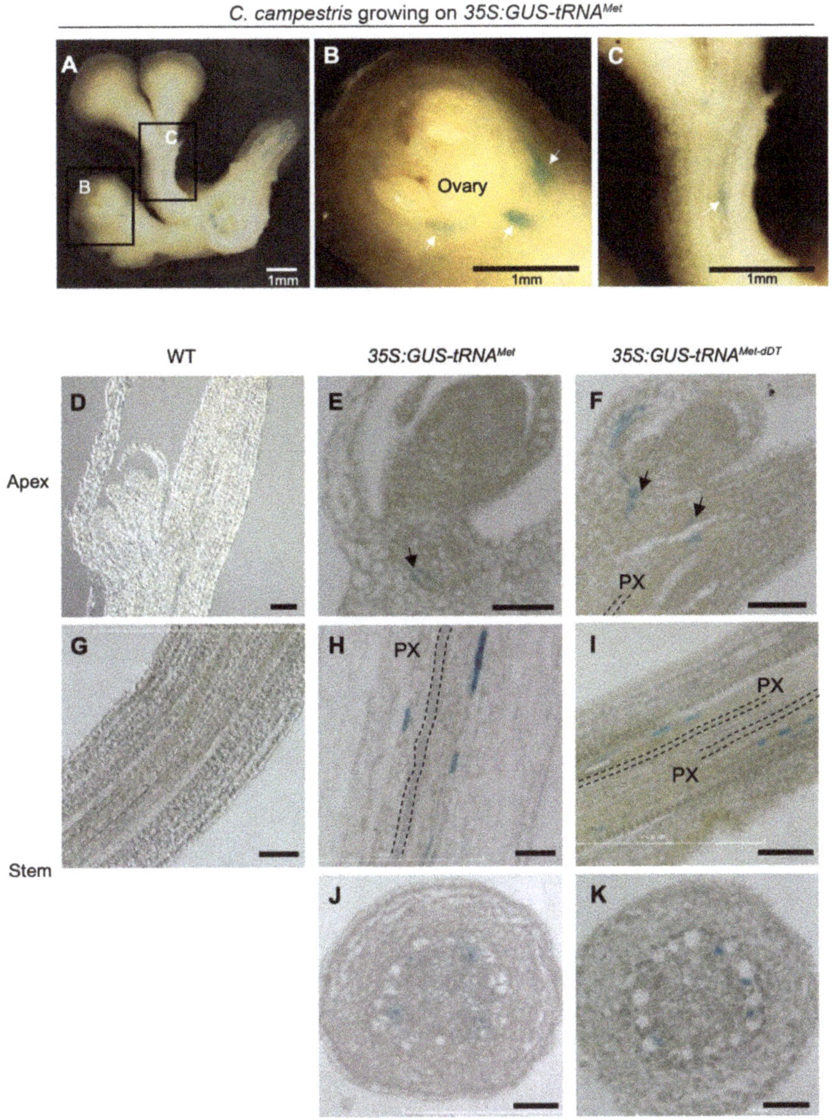

Figure 3. Localization of GUS activity in *C. campestris* stems on *Arabidopsis*. *35S:GUS-tRNA^{Met}* (**A–C,E,H,J**), *35S:GUS-tRNA^{Met-dDT}* (**F,I,K**), and wild type (WT) (**D,G**). (**A–C**) Longitudinally sectioned *C. campestris* flowers from plants on a *35S:GUS-tRNA^{Met}* host. (**B,C**) High-magnification images of (A). White arrows indicate GUS signals in flower and peduncle. (**D–F**) *C. campestris* apices (segment 1–12 cm from host) and (**G–K**) stems (segment 0–2 cm) were GUS stained, embedded in paraffin, and (**G,I**) longitudinally, or (**J,K**) transversely, sectioned in 20 μm-thickness. Black arrows indicate the apical termini of sieve tube. PX, parasite xylem. Scale bars: 100 μm.

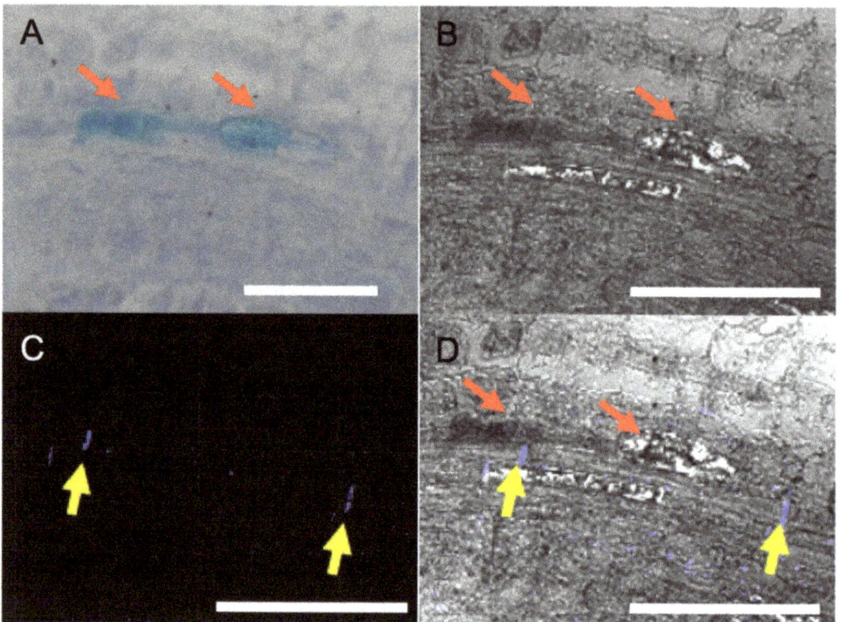

Figure 4. GUS activity detected in the cells adjacent to the aniline-blue-stained sieve tube. A 20 μm-thick paraffin section of *Cuscuta campestris* stem on an *Arabidopsis 35S:GUS-tRNA^{Met}* host was stained with X-gluc for 24 h and aniline blue for 45 min. (**A**) Bright field image by upright microscope. (**B**) Transmission image by confocal laser scanning microscopy. (**C**) Fluorescent image of aniline blue-stained sieve plates by confocal laser scanning microscopy. (**D**) Overlay image of (**B**,**C**). GUS activity (red arrows) was detected in the cells adjacent to the aniline-blue-stained sieve tube (yellow arrows). Scale bar: 100 μm.

3. Discussion

The fate and function of mobile mRNAs in plants has been the subject of speculation and research since the earliest reports of systemically trafficked mRNAs in plants [26,27]. These issues are all the more intriguing when they occur in the context of host-parasite trans-species interactions. Recent breakthroughs have contributed to understanding how the mobility of mRNAs is regulated in plants and have shown that mobile mRNAs may be translated into proteins in their destination cells [19,20], but the subject has yet to be resolved in parasitic plant interactions. We used TLS-mediated mRNA mobility to simultaneously investigate mechanisms regulating mRNA transfer and translation of the mRNA in *C. campestris* feeding on transgenic plants.

The fusion of tRNA sequences to the *GUS* gene conferred mobility on *GUS* mRNA from *Arabidopsis* into attached *C. campestris* (Figure 2; Supplementary Figure S1). Subsequent translation to protein resulted in consistent detection of GUS enzyme activity in these *C. campestris* shoots (Figures 1 and 2). Our results were consistent for two *tRNA^{Ile}* constructs (with or without an ER localization signal peptide) and independently verified in two different laboratories (Japan and the U.S.A.). These data confirm a lack of mobility for *GUS* encoded by constructs missing the tRNAs or for *GUS* fused to *tRNA$^{Met-dAT}$* in host-parasite systems. Our findings are largely consistent with the graft transmissibility of *GUS-tRNA* fusions reported by Zhang et al. [20], who also demonstrated the mRNA mobility of *GUS* fused to *tRNA^{Met}*, *tRNA$^{Met-dDT}$*, and *tRNA^{Gly}* (compare to Table 1). One discrepancy between the *Arabidopsis* graft studies and our host-*Cuscuta* data is the mobility of *GUS-tRNA^{Ile}* into *C. campestris*, whereas no graft transmissibility of this tRNA fusion was seen [28]. Taken together, these results suggest that the regulation of mRNA movement across the *C. campestris* haustorial connection is similar, but not identical to, an *Arabidopsis* graft junction.

The presence of a TLS element associated with mRNA is just one of the mechanisms currently known to facilitate phloem mobility, but we wondered whether it could account for the large number of mobile mRNAs in *C. campestris* parasitizing *Arabidopsis*. To test this, we evaluated 492 of the most abundant mobile *Arabidopsis* mRNAs from a list of nearly 8000 previously reported host-to-*Cuscuta* mobile mRNAs [17]. Of these genes, 392 (79.6%) are reported as also being cell-to-cell mobile mRNAs in *Arabidopsis* (www.arabidopsis.org). We searched these 392 genes for a TLS structure and found that 35 genes (8.9%) had a TLS. This is consistent with a previous report that 11.4% of *Arabidopsis* mobile mRNAs identified from a grafting study have a TLS [20]. We conclude that the TLS motif is likely just one of several mechanisms to regulate host-*Cuscuta* mobility of mRNAs [19,29], yet this is an important finding in that it illustrates a simple mechanism for engineering mRNA mobility in a gene that otherwise may not be mobile. This will be a useful experimental tool for further investigations of host-*Cuscuta* interactions.

Cuscuta spp. are known to take proteins directly from their hosts. This has been shown for phloem-expressed, soluble GFP [30,31] and phosphinothricin acetyl transferase [32]. Recently, large-scale movement of proteins from *Arabidopsis* and soybean hosts to *C. australis* has been described, including direct mobility of a GUS protein [3]. This stands in contrast to our work in which no evidence of GUS protein movement was detected. The work with *C. australis* did not include extra sequences with the *GUS* gene construct to enhance mobility, and the case for mobility was made based on detection of GUS activity in the absence of successful amplification of *GUS* mRNA from the same tissues. It is difficult to reconcile the difference in our two studies, although slightly different methodologies were used. The simplest answer may lie in potential differences in haustorial function between *C. campestris* and *C. australis*, and this subject warrants further investigation. It is likely that both mechanisms operate, and Liu et al. [3] concede that in their system some amount of host-encoded protein may arrive in the parasite through the translation of mobile mRNA. The larger question may revolve around the relative contributions of direct movement of mature proteins as compared to mRNA intermediates.

Localization of GUS expression in the parasite suggests that GUS mRNA moves long distances in the parasite and is imported into companion cells or phloem parenchyma cells of *C. campestris* (Figures 3 and 4). The GUS activity was observed near shoot apices and floral organs, although it was not detected inside these structures. The pattern of staining of specific cells or groups of cells may be an artifact of the sectioning and staining methodology, or may reflect the uptake and translation of mobile mRNAs by specific cells, as suggested by targeted the synthesis and translation of mobile mRNA in specific phloem companion cells [33,34].

Taken together, the appearance of functional host protein in the parasite raises intriguing possibilities for novel organismal interactions. There is little doubt that direct protein exchange occurs between parasitic plants and their hosts, but mobile mRNAs encoding proteins that are membrane bound or too large to easily translocate would provide another avenue of plant-plant interaction. Just as recent studies of *C. campestris* microRNAs have demonstrated a role for these molecules in suppressing expression of specific host genes [35], mobile mRNAs may provide an additional means of host manipulation. It will be interesting to investigate the functional significance of this process.

4. Material and Methods

4.1. Plant Material and Growth Conditions

Experiments were conducted in two locations, with consistent results despite minor differences in growth conditions. In Japan, *C. campestris* seeds were harvested from lab-grown plants parasitizing *Nicotiana tabacum* hosts grown at 25 °C with 16 h light and

8 h dark cycles. Experimental growth conditions of *C. campestris* and *Arabidopsis* were described previously [36]. In the US, seedlings of a lab-growth line of *C. campestris* [17] were inoculated on beets (*Beta vulgaris*) and grown for one month at 25 °C with 14 h light and 10 h dark cycles. Pieces of *C. campestris* shoot tip (around 5 cm long) growing on beets were inoculated on the middle of *Arabidopsis* flowering stems (around 7 cm long). To promote coiling, plants were grown under a 65W Spot-Gro Plant Light (Sylvania) with 14 h light and 10 h dark cycles for two weeks.

Arabidopsis seeds were stratified in water at 4 °C for a day and then sown onto Sungro Professional Growing Mix. Plants were grown in a Conviron (Controlled Environments, Inc.) growth chamber with 9 h light and 15 h dark cycles for 6 to 8 weeks before inoculation with *C. campestris*.

4.2. Arabidopsis Plants Expressing ER-GUS with tRNAs

For cloning endoplasmic reticulum (ER) signal peptides fused to *GUS-tRNA* constructs, gDNAs were first extracted from transgenic *Arabidopsis* lines expressing *GUS-tRNA*^{Met}, *GUS-tRNA*$^{Met-dDT}$, *GUS-tRNA*^{Gly}, and *GUS-tRNA*^{Ile} [20]. These gDNAs were used as templates for cloning to insert the ER signal sequence into *ER-GUS-tRNA* constructs. ER-GUS with different tRNAs were cloned into pEarleyGate100 using the forward primer (with 23 amino acid ER targeting signal peptide from AT1G21270) and tRNA specific reverse primers (Supplemental Table S2) [37]. Transgenic *Arabidopsis* plants were generated by floral dipping [38], and at least five individual T2 lines were tested in this study.

4.3. Histochemical and Quantitative GUS Assays

Haustorial regions of two-week-old *C. campestris* attachments on various *Arabidopsis* transgenic lines were collected and embedded in 5% agarose. Using a VT1200 S fully automated vibrating blade microtome (Leica), agarose blocks with plant tissues were sectioned with 400 μm thickness and 0.8 mm/sec speed. Sliced tissues were collected into 48 well plates for further analysis. For the GUS staining, sectioned samples were stained with X-gluc solution for 2 h and destained in 70% EtOH for 10 h.

4.4. Paraffin Embedding

GUS-stained *Cuscuta* stems were fixed with 4% (*w/v*) paraformaldehyde phosphate buffer solution (FUJIFILM Wako Pure Chemical Corporation, Osaka, Japan) at room temperature for 24 h. Fixed samples were dehydrated and embedded in paraffin (Paraplast, Leica Biosystems, Wetzlar, Germany) as described previously [36]. Paraffin blocks were cut into 20 μm-thick sections by using a microtome (PR-50, Yamato Kohki, Asaka, Japan). Sections were extended with water on MAS-coated slide glass (Matsunami Glass Ind., Ltd., Kishiwada, Japan) and dewaxed as described previously [36]. Samples were observed by a BX53 Upright Microscope (Olympus, Tokyo, Japan, https://www.olympus-lifescience.com/, accessed on 16 November 2021).

For the histochemical GUS staining, sectioned samples were stained by GUS staining solution with 5-bromo-4- chloro-3-indolyl-BD-glucuronide (X-gluc) (Fisher) for 3 h in accordance with the guidelines of the manufacturer and photographed using a stereo-zoom microscope (Discovery V12, Carl Zeiss, Jena, Germany).

The fluorescent β-galactosidase assay with 4-MUG (Fisher) was conducted to detect GUS activity under high liquid treatment. Plant samples from *Arabidopsis* and *C. campestris* were collected, and total proteins were extracted in accordance with the guidelines of the manufacturer. Concentrations of total proteins were quantified by Bradford assay (Bio-Rad, Hercules, CA, USA) using bovine serum albumin (Promega, Madison, WI, USA) as a standard. For the 4-MUG assay, fluorescence was detected at the excitation/emission wavelengths of 365 nm/455 nm by a plate reader machine (Biotek Synergy HT). The GUS enzyme activity was expressed as picomoles of 4-methylumbelliferone (MU) (Sigma) produced per milligram protein per minute. Based on standard curves, the results of the 4-MUG assay were calculated.

4.5. Phloroglucinol-HCl (Wiesner) Staining

Phloroglucinol (3%) (Sigma) dissolved in ethanol was mixed with concentrated HCl (Sigma) to make the phloroglucinol-HCl (Wiesner) staining solution [39]. Sectioned tissues were dipped into the solution for 5 min and directly observed under a stereo-zoom microscope (Discovery V12, Carl Zeiss).

4.6. Reverse Transcriptase (RT) PCR and Quantitative PCR

Total RNAs were extracted from at least five independent biological replicates of *Arabidopsis* or *C. campestris* stems using the Trizol reagent and in accordance with the protocol of the manufacturer (Invitrogen). Equal amounts of extracted total RNAs were reverse transcribed using random primers and M-MLV in accordance with the protocol of the manufacturer (High Capacity cDNA Reverse Transcription Kit, ABI).

Gene-specific primers (Table S2) were used in RT-PCR with iProof High-Fidelity DNA Polymerase (Bio-rad) to amplify genes of interest. *GUS-plus* primers were used to measure the *GUS* mRNA movement from host plants into *C. campestris* stems. *CcActin 8* was a positive control to check the equal amount of RNA.

For quantitative RT-PCR (qRT-PCR), *C. campestris* stems (approximately 12 cm-long from the parasite haustorial site to the apical tip) were divided into six segments (2 cm each). Total RNAs were extracted from three biological replicates of *C. campestris* stems by using RNeasy Plant Mini Kit (QIAGEN, Hilden, Germany). First cDNA strand synthesis was performed by using oligio(dT) primer and ReverTra Ace (TOYOBO, Osaka, Japan). qRT-PCR of GUS transcript was performed by using gene specific primers (Table S2), Fast SYBR™ Green Master Mix (Thermo Fisher Scientific, Waltham, MA, USA), and the StepOnePlus Real-Time PCR System (Thermo Fisher Scientific, https://corporate.thermofisher.com/, accessed on 16 November 2021). Standard curves were generated by using partial GUS sequence cloned in a plasmid pCR™-Blunt II-TOPO® (Thermo Fisher Scientific) as a template.

4.7. Aniline-Blue Staining

Dewaxed paraffin sections, 20 μm-thickness, of *C. campestris* stems were stained for 45 min with 1% (w/v) aniline blue solution dissolved in 50mM $NaPO_4$ buffer, pH 7.0, and washed by sterile water twice. Fluorescence were observed by using BX53 Upright Microscope (Olympus) and a laser-scanning confocal microscope (Leica TCS SP8, Leica Biosystems).

4.8. Searching tRNA-like Structure (TLS) Motif in Mobile Host Genes

A database, containing 492 mRNAs that had been found to be mobile from *Arabidopsis* to *Cuscuta* [17], was screened to determine the presence of a TLS motif. The full-length sequences of each mobile mRNA were obtained from TAIR (arabidopsis.org). PlaMoM (Plant Mobile Macromolecules) (http://www.systembioinfo.org/plamom/, accessed on 16 November 2021) provides a search tool to predict a TLS element [40] and was used to analyze the mobile *Arabidopsis* genes to identify presence of any TLS motif.

5. Conclusions

We have addressed the question of whether host-encoded mRNA could be translated into a functional protein following translocation into the parasitic plant *C. campestris*. As part of this work, we used tRNA gene sequences as signals for long-distance trafficking of mRNAs [20]. We observed that GUS-tRNA fusions expressed in *Arabidopsis* hosts resulted in detection of both GUS mRNA and GUS enzymatic activity in associated *C. campestris* shoots. Furthermore, this GUS expression appeared in *C. campestris* tissues near the haustorial connections as well as in shoots and floral organs located distantly from the point of host attachment. GUS expression was associated with the parasite vascular system, suggesting that mobile mRNAs are translated in companion cells or phloem parenchyma. The fact that functional GUS enzyme was produced in the parasite raises the possibility that mobile mRNAs lead to exchange of proteins that may affect the physiology of one or both plants in the parasite-host interaction. Considering the breadth of diversity in mobile mRNAs [17], it is interesting to consider a potential role for mRNAs in parasitic plant communication.

Supplementary Materials: The following supporting information can be downloaded at: https://www.mdpi.com/article/10.3390/plants11010093/s1, Figure S1: GUS staining and quantification of GUS transcript levels using *Cuscuta* on *35S:GUS-tRNA^Met* and *35S:GUS-tRNA^{Met-dDT}* *Arabidopsis*.; Figure S2: Histochemical localization of β-glucuronidase and xylem.; Figure S3: GUS activity detected in the cells adjacent to aniline-blue-stained sieve tube.; Table S1: β-Glucosidase activities in host *Arabidopsis* plants.; Table S2: List of primers used in this study.

Author Contributions: K.A. and J.H.W. independently conceived the experiments; S.-Y.P. and K.S. conducted assays of GUS activity and mRNA movement; S.-Y.P. generated ER-tagged constructs and plants; K.S. conducted high resolution GUS staining experiments; J.B. analyzed TLS sequences and provided technical support; S.-Y.P. and K.S. analyzed the data; S.-Y.P. drafted the manuscript with additional help from J.H.W., K.A. and K.S.; K.A. and J.H.W. share equal responsibility and agree to serve as co-corresponding authors. All authors have read and agreed to the published version of the manuscript.

Funding: Financial support was provided by US National Science Foundation award IOS-1645027 and US Department of Agriculture National Institute of Food and Agriculture award (VA-160111) to J.H.W.; Grant-in-Aid for JSPS Fellows (19J14848, JSPS) to K.S.; and Grant-in-Aid for Scientific Research (18H03950 and 19H00944, JSPS) to K.A.

Institutional Review Board Statement: Not applicable.

Informed Consent Statement: Not applicable.

Acknowledgments: We thank Friedrich Kragler (Max Planck Institute) for GUS-tRNA fusion plants and Junpei Takano (Osaka Prefecture University) for the help in confocal laser scanning microscopy.

Conflicts of Interest: The authors declare no conflict of interest.

References

1. Costea, M.; Tardif, F.J. The biology of Canadian weeds. 133. *Cuscuta campestris* Yuncker, *C. gronovii* Willd. ex Schult., *C. umbrosa* Beyr. ex Hook., *C. epithymum* (L.) L. and *C. epilinum* Weihe. *Can. J. Plant Sci.* **2006**, *86*, 293–316. [CrossRef]
2. Shimizu, K.; Aoki, K. Development of parasitic organs of a stem holoparasitic plant in genus Cuscuta. *Front. Plant Sci.* **2019**, *10*, 11. [CrossRef]
3. Liu, N.; Shen, G.; Xu, Y.; Liu, H.; Zhang, J.; Li, S.; Li, J.; Zhang, C.; Qi, J.; Wang, L.; et al. Extensive inter-plant protein transfer between Cuscuta parasites and their host plants. *Mol. Plant* **2019**, *13*, 573–585. [CrossRef]
4. Kim, G.; Westwood, J.H. Macromolecule exchange in *Cuscuta*–host plant interactions. *Curr. Opin. Plant Biol.* **2015**, *26*, 20–25. [CrossRef] [PubMed]
5. Johnson, N.R.; Axtell, M.J. Small RNA warfare: Exploring origins and function of trans-species microRNAs from the parasitic plant *Cuscuta*. *Curr. Opin. Plant Biol.* **2019**, *50*, 76–81. [CrossRef]
6. Yang, Z.; Wafula, E.K.; Kim, G.; Shahid, S.; McNeal, J.R.; Ralph, P.E.; Timilsena, P.R.; Yu, W.-B.; Kelly, E.A.; Zhang, H.; et al. Convergent horizontal gene transfer and cross-talk of mobile nucleic acids in parasitic plants. *Nat. Plants* **2019**, *2019*. *5*, 991–1001. [CrossRef]
7. Kehr, J.; Buhtz, A. Long distance transport and movement of RNA through the phloem. *J. Exp. Bot.* **2007**, *59*, 85–92. [CrossRef]
8. Lu, K.-J.; Huang, N.-C.; Liu, Y.-S.; Lu, C.-A.; Yu, T.-S. Long-distance movement of Arabidopsis FLOWERING LOCUS T RNA participates in systemic floral regulation. *RNA Biol.* **2012**, *9*, 653–662. [CrossRef]
9. Kim, M.; Canio, W.; Kessler, S.; Sinha, N. Developmental Changes Due to Long-Distance Movement of a Homeobox Fusion Transcript in Tomato. *Science* **2001**, *293*, 287–289. [CrossRef]
10. Banerjee, A.K.; Chatterjee, M.; Yu, Y.; Suh, S.-G.; Miller, W.A.; Hannapel, D.J. Dynamics of a mobile RNA of potato involved in a long-distance signaling pathway. *Plant Cell* **2006**, *18*, 3443–3457. [CrossRef]
11. Haywood, V.; Yu, T.-S.; Huang, N.-C.; Lucas, W.J. Phloem long-distance trafficking of *GIBBERELLIC ACID-INSENSITIVE* RNA regulates leaf development. *Plant J.* **2005**, *42*, 49–68. [CrossRef] [PubMed]
12. Hannapel, D.J. Long-Distance Signaling via Mobile RNAs. In *Long-Distance Systemic Signaling and Communication in Plants*; Baluška, F., Ed.; Springer: Berlin/Heidelberg, Germany, 2013; pp. 53–70.
13. Kehr, J.; Kragler, F. Long distance RNA movement. *New Phytol.* **2018**, *218*, 29–40. [CrossRef]
14. Thieme, C.J.; Rojas-Triana, M.; Stecyk, E.; Schudoma, C.; Zhang, W.; Yang, L.; Miñambres, M.; Walther, D.; Schulze, W.X.; Paz-Ares, J.; et al. Endogenous *Arabidopsis* messenger RNAs transported to distant tissues. *Nat. Plants* **2015**, *1*, 15025. [CrossRef]

15. Yang, Y.Z.; Mao, L.Y.; Jittayasothorn, Y.; Kang, Y.M.; Jiao, C.; Fei, Z.J.; Zhong, G.-Y. Messenger RNA exchange between scions and rootstocks in grafted grapevines. *BMC Plant Biol.* **2015**, *15*, 251. [CrossRef] [PubMed]
16. Lu, X.H.; Liu, W.Q.; Wang, T.; Zhang, J.L.; Li, X.J.; Zhang, W.N. Systemic long-distance signaling and communication between rootstock and scion in grafted vegetables. *Front. Plant Sci.* **2020**, *11*, 460. [CrossRef]
17. Kim, G.; LeBlanc, M.L.; Wafula, E.K.; de Pamphilis, C.W.; Westwood, J.H. Genomic-scale exchange of mRNA between a parasitic plant and its hosts. *Science* **2014**, *345*, 808–811. [CrossRef] [PubMed]
18. Banerjee, A.K.; Lin, T.; Hannapel, D.J. Untranslated regions of a mobile transcript mediate RNA metabolism. *Plant Physiol.* **2009**, *151*, 1831–1843. [CrossRef]
19. Yang, L.; Perrera, V.; Saplaoura, E.; Apelt, F.; Bahin, M.; Kramdi, A.; Olas, J.; Mueller-Roeber, B.; Sokolowska, E.; Zhang, W.; et al. m5C methylation guides systemic transport of messenger RNA over graft junctions in plants. *Curr. Biol.* **2019**, *29*, 2465–2476.e5. [CrossRef] [PubMed]
20. Zhang, W.; Thieme, C.J.; Kollwig, G.; Apelt, F.; Yang, L.; Winter, N.; Andresen, N.; Walther, D.; Kragler, F. tRNA-Related Sequences Trigger Systemic mRNA Transport in Plants. *Plant Cell* **2016**, *28*, 1237–1249. [CrossRef]
21. Schoenbeck, M.A.; Swanson, G.A.; Brommer, S.J. β-Glucuronidase activity in seedlings of the parasitic angiosperm *Cusctua pentagona*: Developmental impact of the β-glucuronidase inhibitor saccharic acid 1,4-lactone. *Funct. Plant Biol.* **2007**, *34*, 811–821. [CrossRef]
22. Waigmann, E.; Zambryski, P. Tobacco mosaic virus movement protein-mediated protein transport between trichome cells. *Plant Cell* **1995**, *7*, 2069–2079.
23. Crawford, K.M.; Zambryski, P.C. Subcellular localization determines the availability of non-targeted proteins to plasmodesmatal transport. *Curr. Biol.* **2000**, *10*, 1032–1040. [CrossRef]
24. Stadler, R.; Wright, K.M.; Lauterbach, C.; Amon, G.; Gahrtz, M.; Feuerstein, A.; Oparka, K.J.; Sauer, N. Expression of GFP-fusions in *Arabidopsis* companion cells reveals non-specific protein trafficking into sieve elements and identifies a novel post-phloem domain in roots. *Plant J.* **2005**, *41*, 319–331. [CrossRef] [PubMed]
25. Liljegren, S. Phloroglucinol stain for lignin. *Cold Spring Harb. Protoc.* **2010**, *2010*, pdb.prot4954. [CrossRef]
26. Ruiz-Medrano, R.; Xoconostle-Cazares, B.; Lucas, W. Phloem long-distance transport of CmNACP mRNA: Implications for supracellular regulation in plants. *Development* **1999**, *126*, 4405–4419. [CrossRef]
27. Lucas, W.J.; Yoo, B.-C.; Kragler, F. RNA as a long-distance information macromolecule in plants. *Nat. Rev. Mol. Cell Biol.* **2001**, *2*, 849–857. [CrossRef] [PubMed]
28. Zhang, S.; Sun, L.; Kragler, F. The phloem-delivered RNA pool contains small noncoding RNAs and interferes withtranslation. *Plant Physiol.* **2009**, *150*, 378–387. [CrossRef] [PubMed]
29. Calderwood, A.; Kopriva, S.; Morris, R.J. Transcript abundance axplains mRNA mobility data in *Arabidopsis thaliana*. *Plant Cell* **2016**, *28*, 610–615. [CrossRef] [PubMed]
30. Haupt, S.; Oparka, K.J.; Sauer, N.; Neumann, S. Macromolecular trafficking between *Nicotiana tabacum* and the holoparasite *Cuscuta reflexa*. *J. Exp. Bot.* **2001**, *52*, 173–177. [CrossRef] [PubMed]
31. Birschwilks, M.; Haupt, S.; Hofius, D.; Neumann, S. Transfer of phloem-mobile substances from the host plants to the holoparasite *Cuscuta* sp. *J. Exp. Bot.* **2006**, *57*, 911–921. [CrossRef] [PubMed]
32. Jiang, L.; Qu, F.; Li, Z.; Doohan, D. Inter-species protein trafficking endows dodder (*Cuscuta pentagona*) with a host-specific herbicide-tolerant trait. *New Phytol.* **2013**, *198*, 1017–1022. [CrossRef]
33. Chen, Q.; Payyavula, R.S.; Chen, L.; Zhang, J.; Zhang, C.; Turgeon, R. FLOWERING LOCUS T mRNA is synthesized in specialized companion cells in *Arabidopsis* and Maryland Mammoth tobacco leaf veins. *Proc. Natl. Acad. Sci. USA* **2018**, *115*, 2830–2835. [CrossRef]
34. Slewinski, T.L.; Zhang, C.; Turgeon, R. Structural and functional heterogeneity in phloem loading and transport. *Front. Plant Sci.* **2013**, *4*, 244. [CrossRef]
35. Shahid, S.; Kim, G.; Johnson, N.R.; Wafula, E.; Wang, F.; Coruh, C.; Bernal-Galeano, V.; Phifer, T.; de Pamphilis, C.W.; Westwood, J.H.; et al. MicroRNAs from the parasitic plant *Cuscuta campestris* target host messenger RNAs. *Nature* **2018**, *553*, 82–85. [CrossRef]
36. Hozumi, A.; Bera, S.; Fujiwara, D.; Obayashi, T.; Yokoyama, R.; Nishitani, K.; Aoki, K. Arabinogalactan proteins accumulate in the cell walls of searching hyphae of the stem parasitic plants, *Cuscuta campestris* and *Cuscuta japonica*. *Plant Cell Physiol.* **2017**, *58*, 1868–1877. [CrossRef]
37. Lao, J.; Oikawa, A.; Bromley, J.R.; McInerney, P.; Suttangkakul, A.; Smith-Moritz, A.M.; Plahar, H.; Chiu, T.-Y.; Fernández-Niño, S.M.G.; Ebert, B.; et al. The plant glycosyltransferase clone collection for functional genomics. *Plant J.* **2014**, *79*, 517–529. [CrossRef]
38. Clough, S.J.; Bent, A.F. Floral dip: A simplified method for *Agrobacterium*-mediated transformation of *Arabidopsis thaliana*. *Plant J.* **1998**, *16*, 735–743. [CrossRef]
39. Mitra, P.P.; Loqué, D. Histochemical Staining of Arabidopsis thaliana Secondary Cell Wall Elements. *J. Vis. Exp. JoVE* **2014**, *87*, 51381. [CrossRef]
40. Guan, D.; Yan, B.; Thieme, C.; Hua, J.; Zhu, H.; Boheler, K.R.; Zhao, Z.; Kragler, F.; Xia, Y.; Zhang, S. PlaMoM: A comprehensive database compiles plant mobile macromolecules. *Nucleic. Acids Res.* **2017**, *45*, D1021–D1028. [CrossRef]

Article

Evaluating Branched Broomrape (*Phelipanche ramosa*) Management Strategies in California Processing Tomato (*Solanum lycopersicum*)

Matthew J. Fatino * and Bradley D. Hanson

Department of Plant Sciences, University of California, Davis, CA 95616, USA; bhanson@ucdavis.edu
* Correspondence: mfatino@ucdavis.edu

Citation: Fatino, M.J.; Hanson, B.D. Evaluating Branched Broomrape (*Phelipanche ramosa*) Management Strategies in California Processing Tomato (*Solanum lycopersicum*). *Plants* **2022**, *11*, 438. https://doi.org/10.3390/plants11030438

Academic Editors: Evgenia Dor and Yaakov Goldwasser

Received: 1 December 2021
Accepted: 1 February 2022
Published: 5 February 2022

Publisher's Note: MDPI stays neutral with regard to jurisdictional claims in published maps and institutional affiliations.

Abstract: Detections of the regulated noxious parasitic weed branched broomrape (*Phelipanche ramosa*) in California tomato fields have led to interest in eradication, sanitation, and management practices. Researchers in Israel developed a decision-support system and herbicide treatment regime for management of Egyptian broomrape (*P. aegyptiaca*) in tomato. Research was conducted in 2019 and 2020 to evaluate whether similar treatments could be used to manage branched broomrape in California processing tomatoes and to provide registration support data for the herbicide use pattern. Treatment programs based on preplant incorporated (PPI) sulfosulfuron and chemigated imazapic were evaluated in 2019 and 2020 to determine safety on the processing tomato crop and on common rotational crops. Three single-season tomato safety experiments were conducted and a single rotational crop study was conducted in which a tomato crop received herbicide treatments in 2019 and several common rotational crops were planted and evaluated in 2020 in a site without branched broomrape. In 2020, an efficacy study was conducted in a commercial tomato field known to be infested with branched broomrape to evaluate the efficacy of PPI sulfosulfuron and chemigated imazapic, imazapyr, imazethapyr, and imazamox. After two field seasons, sulfosulfuron and imazapic appeared to have reasonable crop safety on tomato in California; however, rotational crop restrictions will need to be considered if sulfosulfuron is used to manage branched broomrape. In the efficacy study, there was a trend in which the sulfosulfuron and imidazolinone treatments had fewer broomrape shoots per plot than the grower standard treatments, however, none were fully effective and there were no significant differences among the various sulfosulfuron and imidazolinone treatment combinations. Additional research is needed to optimize the treatment timing for management of branched broomrape in this cropping system. Because of registration barriers with imazapic in the California market, future research will focus on treatment combinations of PPI sulfosulfuron and chemigated imazamox rather than imazapic.

Keywords: chemigation; crop safety; branched broomrape; imazapic; imazamox; parasitic plants; sulfosulfuron; weed control

1. Introduction

Processing tomato is an important cash crop to annual agricultural systems in the Central Valley of California. In 2020, California produced 11.4 million tons of tomatoes on 93,000 hectares making up over 95% of US tomato production [1]. Processing tomatoes have a farm-gate value of $1.17 billion and were the 10th most valuable agricultural commodity produced in the state in 2020 [2,3]. California is also important at the international scale, producing about 30% of the world's processing tomatoes [1].

Branched broomrape (*Phelipanche ramosa* syn. *Orobanche ramosa*) is a parasitic plant native to the Mediterranean region of Eurasia. Broomrapes are obligate parasites, lacking chlorophyll, thus obtaining all of their nutrients from parasitized host plants [4]. Broomrape parasitism can substantially reduce the productivity of crop plants, with reproductive tissue disproportionately affected [5].

In the past several years, branched broomrape and Egyptian broomrape (*Phelipanche aegyptiaca*) have been reported in California, including Yolo, Solano, and San Joaquin counties [6]. In California, branched broomrape is an "A" classified pest, being "an organism of known economic importance subject to California State enforced action involving eradication, quarantine regulation, containment, rejection, or other holding action", while Egyptian broomrape is classified as a "Q-listed" noxious weed (having "A-listed" classification pending permanent state determination) [7]. A field reported to be infested with an "A-listed" pest such as branched broomrape will be evaluated by the local county agriculture commissioner, quarantined, and that season's crop destructed. For at least two years following this discovery, a hold order is placed on the field and only approved non-host rotational crops may be planted. Broomrape has been discovered in conventional, intensely managed processing tomato fields, suggesting that conventional weed control practices and currently registered herbicides do not provide adequate broomrape control. Currently, there are no registered management practices that can selectively control branched broomrape, making this parasitic weed a serious threat to the California's processing tomato industry.

Researchers in Israel have developed a decision-support system, named PICKIT, to manage Egyptian broomrape in processing tomatoes [8]. The PICKIT system utilizes two acetolactate synthase (ALS)-inhibiting herbicides to control broomrape; preplant-incorporated (PPI) sulfosulfuron followed by low dose chemigation or foliar applications of imazapic during the growing season. The PPI and chemigation treatments reduce attachment and growth of the parasite while the late season foliar imazapic treatment can be used as a clean-up treatment to kill seeds from emerged broomrape plants or under low infestation conditions [8]. In 2016, commercial tomato growers in Israel deployed the PICKIT system and achieved 95% Egyptian broomrape control in 33 fields [8].

While branched broomrape is currently an "A-list" quarantine pest in California requiring crop destruction, this pest could become widespread enough to require management programs like any other weed. The PICKIT system developed in Israel could provide similar management in California. Because there are differences between the Israeli and California processing tomato systems (climate, irrigation, soil type, crop rotations, variety, etc.) and broomrape species (branched vs. Egyptian), the PICKIT program must be evaluated and calibrated for use in California cropping systems. Sulfosulfuron is registered in many U.S. states for use as a selective systemic herbicide on broadleaf weeds in wheat (*Triticum aestivum*) and is registered in California for non-crop use but not in tomato [9]. Imazapic is registered in the southern United States for use as an early post-emergence herbicide in peanut (*Arachis hypogea*) and for rangeland weed control in much of the U.S. but is not registered in California for any use [10]. In order for these herbicides to potentially be registered under an emergency use authorization or an indemnified label for broomrape control under California production conditions, there must be research on their performance and crop safety. The overall goal of this research was to evaluate treatments based on the PICKIT decision-support system for branched broomrape control in processing tomatoes and to provide registration support data for these herbicides in California.

2. Results

2.1. Crop Safety Evaluations

In the two 2019 crop safety experiments, there were no treatment-related differences in phytotoxicity on processing tomato among treatments (data not shown, [11]). Tomato yield ranged from 16 to 24 kg/m^2 in experiment 1 and 18 to 24 kg/m^2 in experiment 2 (Table 1) and there were no significant differences in tomato yield among treatments in either experiment (p = 0.56, 0.69). Similarly, in the 2020 crop safety experiment, there was no phytotoxicity or height reduction observed on processing tomato in any of the treatment plots (data not shown, [11]) and there were no differences in tomato yield (Table 1).

Table 1. Tomato yield from tomato crop safety experiments conducted in 2019 and 2020 in Yolo County, CA, USA.

Trt [1]	Treatment	Rate	Growing Degree Days (GDD)	5 April 2019		30 May 2019		22 April 2020	
		g ai/ha		Yield (kg/m^2)	SE	Yield (kg/m^2)	SE	Yield (kg/m^2)	SE
1	Control	na	na	20.2	4.1	21.2	1.6	20.2	1.8
2	Control 2 [2]	na	na	24.3	3.4	20.7	2.6	17.5	5.5
3	Sulfosulfuron Imazapic	37.5 4.8	na 400, 500, 600, 700, 800	21.1	0.7	22.1	1.8	17.7	1.5
4	Sulfosulfuron Imazapic	37.5 4.8	na 400, 600	16.8	3.6	18.4	4.5	21.3	5.6
5	Imazapic	4.8	na	17.9	3.7	21.5	2.9	19.0	6.1
6	Sulfosulfuron Imazapic	70 9.6	na 400, 600	21.1	2.3	22.9	2.5	19.9	3.7
7	Sulfosulfuron Imazapic	70 9.6	na 400, 500, 600, 700, 800	21.1	2.3	21.3	3.8	19.6	5.9
8	Imazapic	9.6	na	20.1	3.3	22.4	4.3	17.0	2.9
			p-Value (alpha = 0.05)	0.56		0.69		0.65	

[1] Trt = treatment, ai = active ingredient. [2] Treatment 2 was a placeholder for a commercial standard PRE tank mix that was not applied in any of the experiments. ai = active ingredient. Means separated with one-way analysis of variance followed by Tukey-HSD test in *agricolae* package in R. *n* = 4.

2.2. Rotational Crop Safety Evaluations

In the 2019 tomato crop, there was no treatment-related phytotoxicity (data not shown, [11]) or differences in tomato yield (Table 2). In the 2020 season, there was no phytotoxicity observed in fall-planted wheat (data not shown). There were no differences in height or fresh weight among treatments for sunflower (*Helianthus annuus*), safflower (*Carthamus tinctorium*), or kidney bean (*Phaseolus vulgaris*) (Tables 3 and 4). Cantaloupe (*Cucumis melo* var. *cantalupo*) biomass tended to be lowest following the sulfosulfuron treatments; however, due to plot variability, field bindweed (*Convolvulus arvensis*) competition, and gopher (*Thomomys bottae*) damage, and an application of rimsulfuron before planting (not registered on cantaloupe), these differences cannot be definitively attributed to the experimental herbicide treatments (Tables 3 and 4). Corn (*Zea mays*) planted after sulfosulfuron at 37.5 g ai/ha and 70 g ai/ha rates had lower fresh biomass than control treatments ($p \leq 0.001$), as well as appearing stunted and chlorotic at all three rates (18.75, 37.5, 70 g ai/ha) (Tables 3 and 4).

Table 2. Effects of herbicide treatments on a 2019 processing tomato yield as a part of a rotational crop study conducted in Yolo County, CA, USA.

Trt [1]	Treatment	Rate g (ai/ha)	Application	GDD	Tomato Yield [2] (kg/m^2)	SE
1	Control	na	na	na	20.3	1.5
2	Sulfosulfuron	18.75	PPI	na	20.1	2.1
3	Sulfosulfuron	37.5	PPI	na	18.7	2.0
4	Sulfosulfuron	70	PPI	na	19.3	2.1
5	Imazapic	4.8	CHEM ×5	400, 500, 600, 700, 800	14.7	2.4
6	Imazapic	9.6	CHEM ×5	400, 500, 600, 700, 800	15.6	1.7
7	Imazamox	9.6	CHEM ×5	400, 500, 600, 700, 800	19.9	1.0
8	Imazapyr	9.6	CHEM ×5	400, 500, 600, 700, 800	17.2	1.9
9	Imazethapyr	9.6	CHEM ×5	400, 500, 600, 700, 800	17.2	1.4
	p-Value (alpha = 0.05)				0.31	

[1] Trt = treatment, ai = active ingredient, na = not applicable. [2] Means separated with one-way analysis of variance followed by Tukey-HSD test in *agricolae* package in R. *n* = 4.

Table 3. Mean 2020 rotational crop heights in the season following 2019 herbicide treatments in tomato for management of branched broomrape in California.

Trt [1]	Treatment	Rate	Wheat [2]	Corn	Safflower	Sunflower	Kidney Bean	Cantaloupe
		g (ai/ha)				Height (cm)		
1	Control	na	na	127.2 a	82.0	82.8	37.1	19.5 abc
2	Sulfosulfuron	18.75	na	109.4 ab	85.9	88.1	36.8	17.1 bc
3	Sulfosulfuron	37.5	na	62.1 bc	77.0	84.6	37.6	16.5 c
4	Sulfosulfuron	70	na	45.3 b	81.8	78.2	38.9	11.9 d
5	Imazapic	4.8	na	120.8 a	82.0	82.8	38.6	18.7 abc
6	Imazapic	9.6	na	128.8 a	83.8	82.3	37.3	20.7 abc
7	Imazamox	9.6	na	163.4 a	83.3	91.4	38.4	22.4 a
8	Imazapyr	9.6	na	131.9 a	80.3	81.8	36.3	21.3 ab
9	Imazethapyr	9.6	na	129.1 a	81.5	74.7	39.4	18.0 abc
	p-Value (alpha = 0.05)			<0.001	0.91	0.29	0.86	<0.001
	MSD			54.3	ns	ns	ns	4.6

[1] Trt = treatment, ai = active ingredient, na = not applicable, ns = not significant. [2] Visual crop injury ratings for wheat (chlorosis, stunting) were taken instead of weight (data not shown), and there was no injury observed in any plots. Means followed by the same letter within a column are not statistically different according to Tukey's test ($p < 0.05$).

Table 4. Rotational crop above-ground fresh biomass in 2020 following 2019 herbicide treatments in processing tomato for management of branched broomrape in California.

Trt [1]	Treatment	Rate	Wheat [2]	Corn	Safflower	Sunflower	Kidney Bean	Cantaloupe
		g ai/ha				Fresh Biomass (kg) per Meter of Row [3]		
1	Control	na	na	5.6 a	2.7	6.8	1.2	2.8 a
2	Sulfosulfuron	18.75	na	4.3 ab	3.5	6.5	1.5	1.5 ab
3	Sulfosulfuron	37.5	na	1.4 bc	3.5	5.8	1.3	0.9 ab
4	Sulfosulfuron	70	na	1.1 c	2.8	6.3	1.2	0.2 b
5	Imazapic	4.8	na	5.0 a	3.3	5.9	1.4	2.2 ab
6	Imazapic	9.6	na	5.0 a	3.2	5.7	1.3	2.1 ab
7	Imazamox	9.6	na	6.8 a	3.1	6.1	1.4	2.6 ab
8	Imazapyr	9.6	na	4.7 a	3.2	6.1	1.6	2.2 ab
9	Imazethapyr	9.6	na	5.2 a	3.0	6.2	1.5	2.3 ab
	p-Value (alpha = 0.05)			<0.001	0.69	0.88	0.85	0.03
	MSD			3.1	ns	ns	ns	2.5

[1] Trt = treatment, ai = active ingredient. [2] Visual crop injury ratings for wheat (chlorosis, stunting) were taken instead of weight (data not shown), and there was no injury observed in any plots. [3] Data were analyzed using a one-way analysis of variance and Tukey-HSD in the *agricolae* package in R. Means followed by the same letter within a column are not statistically different according to Tukey's test ($p < 0.05$). MSD = minimum significant difference, ns = not significant, $n = 4$.

2.3. Efficacy Evaluation

Branched broomrape emergence was first observed in late May of 2020 and continued steadily until the termination of the experiment in late July. Individual broomrape cluster numbers per 30-m plot ranged from 0 to 58, with only one plot out of 48 having no broomrape emergence. Broomrape cluster counts from sequential sulfosulfuron and imidazolinone treatments (3, 4, 5, 6, 7, 8, 9, 10, 11) were not significantly different from one another but were numerically lower than other treatments (1, 2, 12) (Table 5). Treatments 6 and 8 upper-limit values of a 3-parameter log-logistic function were significantly lower than all other treatments, while treatments 3 (37.5 g ai/ha sulfosulfuron/4.8 g ai/ha imazapic ×5), 4 (37.5 g ai/ha sulfosulfuron/4.8 g ai/ha imazapic ×2), 5 (4.8 g ai/ha foliar imazapic ×2), 7 (70 g ai/ha sulfosulfuron/9.6 g ai/ha imazapic ×2), 9 (37.5 g ai/ha sulfosulfuron/4.8 g ai/ha imazamox ×5), 10 (37.5 g ai/ha sulfosulfuron/4.8 g ai/ha imazapyr ×5), and 11 (37.5 g ai/ha sulfosulfuron/4.8 g ai/ha imazethapyr ×5) were significantly lower than two of the three other treatments—2 (untreated check 2) and 12 (rimsulfuron). ED50 values from a 3-parameter log-logistic function were not significantly different among treatments which indicates no clear treatment-related acceleration or delay in emergence of branched broomrape (Table 5).

Table 5. Effect of herbicide treatments on broomrape cluster number and predicted value of broomrape emergence in a tomato field trial from a 3-parameter log logistic model using *drc* package in R.

Trt [1]	Treatment Name	Rate g (ai/ha)	Cumulative Broomrape Clusters [2]	b (slope [3]) ± 95 CI	d (upper limit) ± 95 CI	e (ed50) ± 95 CI
1	Control	na	25 ab	-8.5 ± 6.5	20.5 ± 7.5	92.6 ± 11.8
2	Control 2	na	45 a	-12.5 ± 3.4	47.7 ± 4.1	94.0 ± 2.2
3	Sulfosulfuron Imazapic	37.5 4.8	18.3 b	-8.5 ± 6.5	20.5 ± 7.5	92.6 ± 11.8
4	Sulfosulfuron Imazapic	37.5 4.8	13.8 b	-7.9 ± 12.8	15.2 ± 11.4	89.6 ± 25.6
5	Imazapic	4.8	11 b	-7.7 ± 22.3	11.8 ± 12.9	85.3 ± 37.6
6	Sulfosulfuron Imazapic	70 9.6	5.3 b	-13.3 ± 13.1	5.2 ± 1.5	90.4 ± 8.2
7	Sulfosulfuron Imazapic	70 9.6	17.8 b	-14.2 ± 9.1	18.0 ± 3.6	94.3 ± 5.3
8	Imazapic	9.6	7.5 b	-12.3 ± 20.1	7.6 ± 2.5	73.8 ± 12.0
9	Sulfosulfuron Imazamox	37.5 4.8	16.5 b	-10.4 ± 19.5	17.7 ± 12.1	92.4 ± 20.0
10	Sulfosulfuron Imazapyr	37.5 4.8	16.5 b	-7.6 ± 6.8	18.1 ± 6.9	86.2 ± 13.5
11	Sulfosulfuron Imazethapyr	37.5 4.8	15.5 b	-8.4 ± 11.7	17.1 ± 9.7	88.9 ± 18.8
12	Rimsulfuron	43.7	45.3 a	-8.3 ± 4.2	49.9 ± 11.2	90.2 ± 7.4

[1] Trt = treatment, ai = active ingredient, na = not applicable. [2] Data were analyzed using a one-way analysis of variance and Tukey-HSD in the *agricolae* package in R. Means followed by the same letter within a column are not statistically different according to Tukey's test ($p < 0.05$). [3] The slope of the dose-response curve at ED50 has the opposite sign as compared to the sign of the parameter b [12].

3. Discussion

3.1. Crop Safety Evaluations

After two field seasons and three studies, crop safety for sulfosulfuron and imazapic appears acceptable at both the proposed rate structure and two times the proposed rate structure in California processing tomato. These results confirm the crop safety reported for the PICKIT program in Israel. Recent studies have demonstrated that ALS inhibitor herbicides are less injurious to broomrape-parasitized plants compared to unparasitized plants as the parasite acts as a strong sink for herbicides [13]. Sulfosulfuron is registered in California for non-crop use but is not currently registered for use in tomato. Imazapic is not currently registered in California and faces a difficult registration pathway in the state, so future research will focus on another imidazolinone herbicide, imazamox, which has a somewhat more favorable registration pathway. Additional studies will need to be conducted to further evaluate the safety and performance of chemigated imazamox.

3.2. Rotational Crop Safety Evaluation

Based on this initial rotational crop safety experiment, there were few indications of problems related to the imidazolinone herbicides applied five times via chemigation at up to 9.6 g ai/ha (2× of the proposed use rate). There was some early season stunting and chlorosis observed with sulfosulfuron in sunflower, but the plants grew out of this injury. There were some indications of crop safety concerns for PPI sulfosulfuron treatments, primarily for corn and cantaloupe. Seeding across all crops was inconsistent and denser than commercially planted stands. If the herbicides utilized in the PICKIT system are registered in California, tomato growers will have to adjust crop rotations based on the plant-back restrictions associated with sulfosulfuron [9]. Given the importance of tomato in this cropping system, such rotational crop restrictions might be acceptable to growers impacted by branched broomrape. Further research is needed to verify these results and validate the safety and performance of additional imazamox-focused treatment regimes.

3.3. Efficacy Evaluation

Currently, the economic and action threshold for branched broomrape in California is any detection of the parasitic plant. With the exception of a single plot, all of the treatment plots had some broomrape clusters by the end of the season. The sequential sulfosulfuron and imidazolinone treatment plots had fewer broomrape clusters on average than other treatment plots, though the late season foliar-applied treatments (12 June and 25 June) should not have affected early season emergence yet had some of the lowest cumulative number of broomrape clusters (Treatment 8). This is likely due to an uneven distribution of branched broomrape, resulting in some "hot" areas of the field with greater broomrape emergence and "cold" areas with relatively low emergence. The experimental blocking was arranged based on reports of higher broomrape density observed by the grower the previous year; however, this did not completely account for the distribution observed during the experiment and additional experiments are needed to more fully evaluate the efficacy of each individual treatment. Sequential sulfosulfuron and imidazolinone treatments had some effect on broomrape emergence, generally reducing emergence compared to other treatments. However, more studies will need to be conducted to determine the relative efficacy of individual treatments among each other and to further refine rates and treatment protocols to optimize control of branched broomrape in the California production system.

The PICKIT decision-support system is based on a growing degree day (GDD) model developed using Egyptian broomrape. Future research will examine the effects of alternate timing of chemigation treatments to address the temporal difference in development between the two species.

4. Materials and Methods

4.1. Crop Safety Evaluations

Three crop safety studies were conducted in 2019 and 2020 to evaluate the crop safety of the Israeli-developed PICKIT decision-support system on processing tomatoes in California. These studies were conducted at the UC Davis Plant Sciences Field Research Facility near Davis, California (38.539105, 121.783547). The soil composition at this site was 41% sand, 34% silt, and 25% clay with 2.1% OM, 6.98 pH, and estimated CEC of 18.2 $cmol_c$/kg of soil. The site was not infested with branched broomrape; this protocol focused on crop safety. Plots were 12 m long on 1.5 m wide beds with one plant line in the center of the bed. 'Heinz 1662' processing tomato transplants were planted at 30.5 cm spacing. Each 60 m long bed had two 15.9 mm drip lines buried at 30.5 cm with 0.6 L/h emitters spaced every 30.5 cm; one line ran the full length of the beds and was used for crop irrigation and fertigation, the second line was terminated at the end of each plot and connected to an above-ground manifold system which was used to apply the experimental chemigation herbicide treatments. Plots were arranged in a randomized complete block design with four replications per treatment. In 2019, two experiments were conducted to represent two planting dates; an early- to mid-season (25 April) and late-season (30 May) planting and a single early- to mid-season planting (22 April) in 2020.

Preplant incorporated (PPI) applications of sulfosulfuron were made one day before transplanting on 24 April and 29 May 2019 in the early- and late-planted experiments, respectively, and on the day of transplanting, 22 April 2020 (Table 6). Sulfosulfuron was applied using a CO_2 backpack sprayer and three-nozzle boom delivering 280.5 L/ha with TeeJet AIXR 11003 nozzles at 193 kPa (TeeJet Technologies, Wheaton, IL, USA). Sulfosulfuron was mechanically incorporated to 7.6 cm after application, after which tomatoes were mechanically transplanted with a three-row transplanter on 25 April 2019 (early planting), 30 May 2019 (late planting), and 22 April 2020.

Table 6. Growing degree day targets and actual herbicide application dates in crop safety and branched broomrape efficacy studies in processing tomatoes in Yolo County, CA, USA.

Growing Degree Day Target	2019 Crop Safety Early Planting	2019 Crop Safety Late Planting	2020 Crop Safety	2020 Efficacy Study
Preplant-Incorporated (PPI)	24 April	29 May	2 April	27 March
Transplant	25 April	30 May	22 April	30 March
400	5 June	13 June	13 May	2 May
500	7 June	20 June	21 May	8 May
600	11 June	24 June	27 May	14 May
700	13 June	28 June	1 June	22 May
800	20 June	3 July	3 June	26 May
Rimsulfuron (Trt 12 Efficacy)	na	na	na	12 June
Foliar (at est. BR [1] emergence)	16 July	15 August	12 June	12 June [2]
Foliar (approx. 21 days after est. BR emergence)	6 August	6 September	6 July	25 June [2]

[1] BR = broomrape. [2] 12 and 25 June did not coincide with the recommended application timing at broomrape emergence and 21 days after; instead, the first application was made one week after broomrape emergence and the second application was 13 days after that.

The PICKIT system's thermal time model is based on growing degree days, with applications at 400, 500, 600, 700, and 800 GDD after transplanting depending on treatment regimes (Table 7). In 2019, chemigation applications were made through the terminated irrigation line using a 20.8 L/min 12-volt electric pump and 113.5 L tank. Treatments were applied to four plots simultaneously, with a total carrier volume of 96.1 L per treatment resulting in approximately 15.9 L per replicate plot (18.3 m^2). In 2020, chemigation applications were made using CO_2 to inject a chemigation mix into a distribution manifold with valved connections at each plot. Treatments were applied to two replicate plots at once with separate injection ports for replicates 1 and 2 and replicates 3 and 4 to reduce the system volume receiving herbicide-treated water. Herbicides were diluted in 11 L of water and this solution was injected into the already-running irrigation system over approximately 15 min, followed by 20 min of water to flush the distribution lines. Foliar imazapic treatments were made on 16 July 2019, 15 August 2019, and 12 June 2020 and approximately 21 days later (6 August 2019, 6 September 2019, and 6 July 2020) with a CO_2 backpack sprayer and two-nozzle boom delivering 280.5 L/ha with TeeJet AIXR 11005 nozzles at 138 kPa. These applications were made at estimated broomrape emergence and approximately 21 days later, as these studies occurred in uninfested fields. Visual plant phytotoxicity (vigor reduction, stunting, chlorosis) was recorded in all three studies and representative plant height (cm) was recorded in the 2020 study (data not shown; 11). All marketable fruit from one-meter square sections of row were harvested on 4 September 2019, 19 September 2019, and 3 September 2020 at commercial maturity and fresh weights were recorded. Yield data were analyzed using a one-way analysis of variance followed by a Tukey-HSD test using the *agricolae* package in RStudio version 1.2.5033 [12,14].

4.2. Rotational Crop Safety Evaluations

A two-year study was conducted from spring 2019 to fall 2020 to evaluate rotational crop safety of sequential sulfosulfuron and imidazolinone herbicide treatments. This field experiment included a 2019 tomato crop treated with various herbicides (Table 8) followed by a planting of six common rotational crops (wheat, corn, safflower, sunflower, kidney bean, cantaloupe) in 2020. The study was conducted at the UC Davis Department of Plant Sciences Field Research Facility near Davis, California (38.539105, 121.783547).

Table 7. 2019 and 2020 tomato crop safety treatment list.

Trt [1]	Treatment	Application [2]	Rate g (ai/ha)	Application Timing
1	Control	na	na	na
2	Control 2 [2]	na	na	na
3	Sulfosulfuron	PPI	37.5	Before transplant
	Imazapic	CHEM ×5	4.8	400, 500, 600, 700, 800 GDD
4	Sulfosulfuron	PPI	37.5	Before transplant
	Imazapic	CHEM ×2	4.8	400, 600 GDD
5	Imazapic	POST ×2	2.4	BR emergence and approximately 21 days later
6	Sulfosulfuron	PPI	70	Before transplant
	Imazapic	CHEM ×5	9.6	400, 500, 600, 700, 800 GDD
7	Sulfosulfuron	PPI	70	Before transplant
	Imazapic	CHEM ×2	9.6	400, 600 GDD
8	Imazapic	POST ×2	4.8	BR emergence and approximately 21 days later

[1] Trt = treatment, ai = active ingredient, BR = broomrape, GDD = growing degree days, PPI = preplant-incorporated, na = not applicable. [2] Treatment 2 was a placeholder for a commercial standard PRE tank mix that was not applied in any of the experiments; instead, the entire field was treated with 350 g ai/ha S-metolachlor and 91.9 g ai/ha trifluralin.

Table 8. 2019 and 2020 herbicide treatments applied to a tomato crop in a rotational crop safety study in Yolo County, CA, USA.

Trt [1]	Treatment Name	Application [2]	Rate g ai/ha	GDD
1	Control	na	na	na
2	Sulfosulfuron	PPI	18.75	na
3	Sulfosulfuron	PPI	37.5	na
4	Sulfosulfuron	PPI	70	na
5	Imazapic	CHEM ×5	4.8	400, 500, 600, 700, 800
6	Imazapic	CHEM ×5	9.6	400, 500, 600, 700, 800
7	Imazamox	CHEM ×5	9.6	400, 500, 600, 700, 800
8	Imazapyr	CHEM ×5	9.6	400, 500, 600, 700, 800
9	Imazethapyr	CHEM ×5	9.6	400, 500, 600, 700, 800

[1] Trt = treatment, ai = active ingredient, na = not applicable. [2] Application dates in 2019: PPI (5/29), 400 (6/1), 500 (6/25), 600 (7/1), 700 (7/5), 800 (7/15).

The site was not infested with branched broomrape; this experiment focused on crop safety of PPI sulfosulfuron and chemigated imazapic, imazamox, imazapyr, and imazethapyr, none of which are currently registered for use in tomato in the United States. The 2019 tomato main plots were 55 m long on 1.5 m beds with one plant line in the center of the bed. Each bed had one 15.9 mm drip line at a depth of 30.5 cm with 0.6 L/h emitters spaced every 30.5 cm. This drip line was used for crop irrigation and fertigation as well as chemigation treatments. For the 2019 tomato crop, main plots were arranged as whole rows in a randomized complete block design with four replications.

Sulfosulfuron was applied on 29 May 2019 using a CO_2 backpack sprayer and three-nozzle boom delivering 280.5 L/ha with TeeJet AIXR 11003 nozzles at 193 kPa. Sulfosulfuron was mechanically incorporated to 7.6 cm after application. Tomato cultivar 'DRI 319' transplants were planted at a 30.5 cm spacing with a three-row transplanter on 30 May 2019. At each growing degree day target, chemigation applications were made through the drip line using a Venturi-style injection system attached to a cone tank over the course of 45 min, with treatments applied to four replicate plots at once. A single one-meter square section of each plot was harvested on 19 September 2019 and total weight of all fruit was recorded.

Following the tomato harvest in 2019, the tomato crop was destroyed in place with a flail mower. After the crop residue dried, beds were lightly cultivated to reshape beds but minimize soil mixing. The 55 m long tomato main plots were divided into six 9.1 m subplots for the 2020 rotational crops in a split plot design. The six rotational crops included wheat, corn, safflower, sunflower, dry bean, and cantaloupe which were randomly assigned to a subplot such that the 2020 experimental design was a randomized split plot with four replications. On 22 November 2019, wheat subplots were planted with a grain drill. Visual wheat injury (chlorosis, stunting) measurements were recorded during the winter of 2019 and spring of 2020 (data not shown). In mid-April 2020, the experimental area was treated with glyphosate to terminate the wheat and control winter weeds and all plots were lightly cultivated to prepare a seedbed. On 17 April 2020, corn (LG Seeds ES7514), safflower (CW99-OL), sunflower (S.O.C. France, 19044), kidney bean (red kidney), and cantaloupe (Osborne 'Hale's Best Jumbo') were planted using an Earthway precision garden seeder (Earthway Products, Inc., Bristol, IN, USA). Summer crops were irrigated as needed with a single drip irrigation line on the soil surface. Plant height and fresh weight biomass (per 1 m of row) were recorded nine weeks after planting on 23 June 2020; the experiment was subsequently terminated without taking the crops to maturity. Height and fresh biomass data were analyzed using a one-way analysis of variance followed by a Tukey-HSD test with the *agricolae* package in RStudio version 1.2.5033 [12,14].

4.3. Efficacy Evaluation

A study was conducted in a commercial tomato field in Yolo County, CA, USA, which had been reported as infested with branched broomrape in 2019 and a portion of the crop was destroyed under CDFA quarantine provisions. The infested area was prepared for planting by the grower and used for a 2020 experiment to test the efficacy of sequential PPI sulfosulfuron and chemigated or foliar imidazolinone treatments on branched broomrape in California tomato systems. The soil composition at this site was 25% sand, 42% silt, and 33% clay with 2.7% OM, 7.2 pH, and estimated CEC of 23.6 $cmol_c$/kg of soil.

Plots were 30.5 m long on 1.5 m beds with two drip lines—one 22.2 mm drip line buried at 25.4 cm and one 25.4 mm drip line buried at 30.5 cm in the center of the bed. The 25.4 mm line was used for crop irrigation and fertigation of the entire experimental area. The 22.2 mm drip line was terminated at the ends of each plot serving as the dedicated chemigation line with 0.6 L/h emitters at 30.5 cm spacing. Plots were arranged in a randomized complete block design with four replications.

Sulfosulfuron was applied on 27 March 2020 using a CO_2 backpack sprayer and three-nozzle boom delivering 280.5 L/ha with TeeJet AIXR 11003 nozzles at 193 kPa. In addition to the experimental treatments, the entire plot area was treated with S-metolachlor (350 g ai/ha), pendimethalin (87.3 g ai/ha), metribuzin (91.9 g ai/ha), and diazinon (734.9 g ai/ha) on 27 March 2020 by the cooperating grower. The experimental area was mechanically cultivated to incorporate herbicides to 7.6 cm on the same day. Processing tomato cultivar 'BQ271' seedlings were mechanically transplanted on 30 March 2020 with two plant lines in each bed with plants spaced 30.5 cm apart within and between lines. A foliar application of 43.7 g ai/ha rimsulfuron was made by the grower to the entire experimental area after transplanting.

Chemigation applications were made using CO_2 to inject the chemigation mix into a 50.8 mm lay flat hose connected to valved 22.2 mm chemigation lines in each plot. Treatments were applied to two replicate plots at once; plots of the same treatment in replications 1 and 2 and replications 3 and 4 were treated together. Herbicide treatments were mixed in 11 L of solution which was injected into the already-running irrigation system over approximately 15 min, followed by 20 min of water to flush the distribution and chemigation lines. Chemigation applications were made according to the growing degree day schedule in the PICKIT protocol (Table 9). Foliar imazapic treatments were made with a 2-nozzle backpack sprayer delivering 280.5 L/ha with AIXR 11003 nozzles at 193 kPa.

Table 9. Herbicide treatments in a 2020 processing tomato field experiment in Yolo County, CA, USA.

Trt [1]	Treatment	Application	Rate g (ai/ha)	GDD
1	Control	na	na	na
2	Control 2 [2]	na	na	na
3	Sulfosulfuron	PPI	37.5	na
	Imazapic	CHEM ×5	4.8	400, 500, 600, 700, 800
4	Sulfosulfuron	PPI	37.5	na
	Imazapic	CHEM ×2	4.8	400, 600
5	Imazapic	POST ×2	2.4	BR emergence, 21 days later
6	Sulfosulfuron	PPI	37.5	na
	Imazapic	CHEM ×5	9.6	400, 500, 600, 700, 800
7	Sulfosulfuron	PPI	70	na
	Imazapic	CHEM ×2	9.6	400, 600
8	Imazapic	POST ×2	4.8	BR emergence, 21 days later
9	Sulfosulfuron	PPI	37.5	na
	Imazamox	CHEM ×5	4.8	400, 500, 600, 700, 800
10	Sulfosulfuron	PPI	37.5	na
	Imazapyr	CHEM ×5	4.8	400, 500, 600, 700, 800
11	Sulfosulfuron	PPI	37.5	na
	Imazethapyr	CHEM ×5	4.8	400, 500, 600, 700, 800
12	Rimsulfuron	POST	43.7	na

[1] Trt = treatment, ai = active ingredient, na = not applicable, PPI = preplant-incorporated, POST = post-emergence, CHEM = Chemigated, BR = broomrape. [2] Treatment 2 was a placeholder for a planned commercial standard PRE tank mix that ultimately was not applied in the experiment; instead, the entire experimental area was treated with the grower's preplant-incorporated herbicide program of S-metolachlor (350 g ai/ha), pendimethalin (87.3 g ai/ha), metribuzin (91.9 g ai/ha), and diazinon (734.9 g ai/ha) and also with a post-transplant application of 43.7 g ai/ha rimsulfuron.

Broomrape emergence was evaluated three times weekly for seven weeks then once per week for 3 weeks beginning on 1 June 2020. At each evaluation, individual clusters of broomrape shoots were marked with wire construction flags, with different colors representing each week's emergence. Broomrape shoot clusters were counted and recorded weekly. Total broomrape cluster numbers were analyzed using a one-way analysis of variance followed by a Tukey-HSD test in the *agricolae* package in R [12,14]. Broomrape emergence over time was analyzed with a 3-parameter log-logistic function in the *drc* package in RStudio version 1.2.5033 [14,15].

Author Contributions: Conceptualization, B.D.H. and M.J.F.; methodology, B.D.H. and M.J.F.; formal analysis, M.J.F.; investigation, M.J.F.; writing—original draft preparation, M.J.F.; writing—review and editing, B.D.H.; acquisition, B.D.H. All authors have read and agreed to the published version of the manuscript.

Funding: This research was funded by NIFA Minor Crop Pest Management Program Interregional Research Project #4 (IR-4) project IS00330-19-CA02, the California Department of Food and Agriculture Specialty Crop Block Program, grant number 43408, and the California Tomato Research Institute.

Data Availability Statement: The data presented in this study are available in [Fatino, M. Evaluating Branched Broomrape (*Phelipanche ramosa*) Management Strategies in California Processing Tomato (*Solanum lycopersicum*). Master's Thesis, University of California, Davis, CA, USA, 2021].

Acknowledgments: The collaboration and material support provided by Gene Miyao, Schreiner Brothers Farms, Ag Seeds Unlimited, California Transplanting, and Wilbur Ellis Company is gratefully appreciated.

Conflicts of Interest: The authors declare no conflict of interest. The funders had no role in the design of the study; in the collection, analyses, or interpretation of data; in the writing of the manuscript, or in the decision to publish the results.

References

1. United States Department of Agriculture National Agricultural Statistics Service. Available online: https://www.nass.usda.gov/Statistics_by_State/California/Publications/Specialty_and_Other_Releases/Tomatoes/ (accessed on 15 August 2021).
2. Winans, K.; Brodt, S.; Kendall, A. Life cycle assessment of California processing tomato: An evaluation of the effects of evolving practices and technologies over a 10-year (2005–2015) timeframe. *Int. J. Life Cycle Assess* **2019**, *25*, 538–547. [CrossRef]
3. California Agricultural Production Statistics Report. Available online: https://www.cdfa.ca.gov/Statistics/ (accessed on 15 August 2021).
4. Westwood, J.H. The physiology of the established parasite-host association. In *Parasitic Orobanchaeceae*; Joel, D.M., Gressel, J., Musselman, L.J., Eds.; Springer: Berlin/Heidelberg, Germany, 2013; pp. 87–114.
5. Fernández-Aparicio, M.; Reboud, X.; Gibot-Leclerc, S. Broomrape Weeds. Underground Mechanisms of Parasitism and Associated Strategies for their Control: A Review. *Front. Plant Sci.* **2016**, *7*, 135. [CrossRef] [PubMed]
6. Miyao, G. Egyptian broomrape eradication effort in California: A progress report on the joint effort of regulators, university, tomato growers and processors. In Proceedings of the XIV International Symposium on Processing Tomato 1159, Santiago, Chile, 6–9 March 2016; pp. 139–142.
7. Pest Rating Proposals and Final Ratings. Available online: https://blogs.cdfa.ca.gov/Section3162/?p=3853 (accessed on 5 August 2021).
8. Eizenberg, H.; Goldwasser, Y. Control of Egyptian broomrape in processing tomato: A summary of 20 years of research and successful implementation. *Plant Dis.* **2018**, *102*, 1477–1488. [CrossRef] [PubMed]
9. Anonymous. *Cadre Herbicide Label*; EPA Reg. No. 241-364; BASF Corporation: Ludwigshafen, Germany, 2011.
10. Anonymous. *Outrider Herbicide Label*; EPA Reg. No. 59639-223; Valent, U.S.A. Corporation: Walnut Creek, CA, USA, 2016.
11. Fatino, M. Evaluating Branched Broomrape (*Phelipanche ramosa*) Management Strategies in California Processing Tomato (*Solanum lycopersicum*). Master's Thesis, University of California, Davis, CA, USA, 2021.
12. Kniss, A.R.; Streibig, J.C. Statistical Analysis of Agricultural Experiments Using R. Available online: https://Rstats4ag.org (accessed on 19 January 2020).
13. Goldwasser, Y.; Rabinovitz, O.; Gerstl, Z.; Nasser, A.; Paporisch, A.; Kuzikaro, H.; Sibony, M.; Rubin, B. Imazapic Herbigation for Egyptian Broomrape (Phelipanche aegyptiaca) Control in Processing Tomatoes—Laboratory and Greenhouse Studies. *Plants* **2021**, *10*, 1182. [CrossRef] [PubMed]
14. De Mendiburu, F. Agricolae: Statistical Procedures for Agricultural Research. R PACKAGE Verson 1.3-5. 2021. Available online: https://CRAN.R-project.org/package=agricolae (accessed on 11 May 2020).
15. Ritz, C.; Baty, F.; Streibig, J.C.; Gerhard, D. Dose-Response Analysis Using R. *PLoS ONE* **2015**, *10*, e0146021. [CrossRef] [PubMed]

Article

Germination Stimulant Activity of Isothiocyanates on *Phelipanche* spp.

Hinako Miura [1], Ryota Ochi [1], Hisashi Nishiwaki [1], Satoshi Yamauchi [1], Xiaonan Xie [2], Hidemitsu Nakamura [3], Koichi Yoneyama [2] and Kaori Yoneyama [1,4,*

[1] Graduate School of Agriculture, Ehime University, Matsuyama 790-8566, Japan; h652021a@mails.cc.ehime-u.ac.jp (H.M.); ryoutaochi.org.chem@gmail.com (R.O.); nishiwaki.hisashi.mg@ehime-u.ac.jp (H.N.); yamauchi.satoshi.mm@ehime-u.ac.jp (S.Y.)

[2] Center for Bioscience Research and Education, Utsunomiya University, Utsunomiya 321-8505, Japan; xie@cc.utsunomiya-u.ac.jp (X.X.); yone2000@sirius.ocn.ne.jp (K.Y.)

[3] Graduate School of Agriculture and Life Sciences, The University of Tokyo, Tokyo 113-8657, Japan; hidemitsunakamura87@gmail.com

[4] Japan Science and Technology, PRESTO, Kawaguchi 332-0012, Japan

[*] Correspondence: yoneyama.kaori.wx@ehime-u.ac.jp

Abstract: The root parasitic weed broomrapes, *Phelipanche* spp., cause severe damage to agriculture all over the world. They have a special host-dependent lifecycle and their seeds can germinate only when they receive chemical signals released from host roots. Our previous study demonstrated that 2-phenylethyl isothiocyanate is an active germination stimulant for *P. ramosa* in root exudates of oilseed rape. In the present study, 21 commercially available ITCs were examined for *P. ramosa* seed germination stimulation, and some important structural features of ITCs for exhibiting *P. ramosa* seed germination stimulation have been uncovered. Structural optimization of ITC for germination stimulation resulted in ITCs that are highly active to *P. ramosa*. Interestingly, these ITCs induced germination of *P. aegyptiaca* but not *Orobanche minor* or *Striga hermonthica*. *P. aegyptiaca* seeds collected from mature plants parasitizing different hosts responded to these ITCs with different levels of sensitivity. ITCs have the potential to be used as inducers of suicidal germination of *Phelipanche* seeds.

Keywords: germination stimulant; isothiocyanates; *Phelipanche*; structure–activity relationship; suicidal germination

Citation: Miura, H.; Ochi, R.; Nishiwaki, H.; Yamauchi, S.; Xie, X.; Nakamura, H.; Yoneyama, K.; Yoneyama, K. Germination Stimulant Activity of Isothiocyanates on *Phelipanche* spp. *Plants* **2022**, *11*, 606. https://doi.org/10.3390/plants11050606

Academic Editors: Evgenia Dor, Yaakov Goldwasser and Nikolai Ravin

Received: 18 October 2021
Accepted: 15 February 2022
Published: 24 February 2022

Publisher's Note: MDPI stays neutral with regard to jurisdictional claims in published maps and institutional affiliations.

1. Introduction

Broomrapes (*Orobanche* and *Phelipanche* spp.) in the family Orobanchaceae are devastating obligate root parasitic weeds damaging crop production all over the world [1,2]. In general, broomrapes have wide host ranges including legumes, Solanaceae, Asteraceae, and Brassicaceae, etc. The area threatened by broomrapes, as estimated in 1991, is 16 million ha in the Mediterranean and west Asia [3] and is still increasing.

A single root parasitic weed produces up to 100,000 tiny seeds and it is nearly impossible to remove these seeds from infested soils [4]. The seeds of root parasitic weeds can germinate only when they receive host-derived chemical-germination stimulants. This is a sophisticated strategy for survival of the root parasites, since germinated seeds with limited food stock should attach to the host roots within a couple of days, otherwise they will die. Then, only the parasite seeds in the rhizosphere of host plants will germinate by sensing host-derived stimulants [5]. Accordingly, in severely infested soils, one potentially effective control method for root parasitic weeds is to induce seed germination of the parasites in the absence of host plants, termed 'suicidal germination' [6]. In addition, suicidal germination contributes to reduce the parasite seed bank.

Among host-derived germination stimulants for root parasitic weeds, strigolactones (SLs) are the most potent stimulants and are distributed widely in the plant kingdom [5,7].

SLs also act as host recognition signals for symbiotic arbuscular mycorrhizal (AM) fungi in the rhizosphere and are a class of plant hormones that regulate plant architecture and development in plants [4,6]. Strigol was the first isolated SL germination stimulant for *Striga* from root exudates of cotton [8]. So far, more than 30 SLs have been characterized from root exudates of various plant species [9,10]. Importantly, plants exude SLs for AM fungi [11,12] and nitrogen-fixing bacteria [13,14] but not for root parasitic weeds [7,10,15,16]. AM fungi supply mineral nutrients including phosphate and nitrogen to hosts and in turn receive photosynthates from the hosts. More than 80% of land plants form symbiotic relationships with AM fungi [17]. Although AM symbiosis is a reasonable strategy for land plants to effectively obtain mineral nutrients in nutrient limited soils, there are several exceptional plant species which do not form AM symbiosis. The Brassicaceae plants including *Brassica* spp., *Arabidopsis* and winter vegetables like cabbage and broccoli are representative non-mycotrophic plants [17]. As SLs function as plant hormones regulating plant architecture and development, even non-mycotrophic plants produce SLs [10]. In general, host plants of AM fungi enhance SL exudation under nutrient starved conditions but non-hosts do not [15].

Phelipanche ramosa causes severe damage to oilseed rape (*Brassica napus*) in southern France. Since oilseed rape is a non-host of AM fungi, this plant is expected to exude only small amounts of SLs [15]. Thus, it was suggested that oilseed rape plants may exude non-SL germination stimulants. Indeed, our previous study demonstrated that 2-phenylethyl isothiocyanate (ITC) is an active germination stimulant for *P. ramosa* in root exudates of oilseed rape [18]. ITCs are enzymatically formed from glucosinolates which are specific plant secondary metabolites of *Brassica* species while the roles of ITCs in plants are not fully understood [19].

In the present study, various ITCs including 2-phenylethyl ITC were examined for *P. ramosa* seed-germination stimulation to characterize the important structural features for the activity. Some ITCs were found to be highly active to *P. ramosa* and also to *P. aegyptiaca*, inducing germination at as low as 10^{-15} M.

2. Results

2.1. Structure–Activity Relationships of ITCs in Germination Stimulation of Phelipanche *spp.*

Germination stimulation activities of 21 commercially available ITCs (Figure 1A) to *P. ramosa* seeds are shown in Figure 2A. Although alkyl ITCs with hexyl or a shorter alkyl group were very weak stimulants, ITCs with a heptyl (C_7) to dodecyl (C_{12}) but not a tetradecyl group showed strong germination stimulation activities. Benzyl ITC was as active as 2-phenylethyl ITC, while phenyl ITC was inactive.

The 3-phenylpropyl, 4-phenylbutyl, 5-phenylpentyl, and 6-phenylhexyl ITCs were prepared and examined for their germination stimulation activity on *P. ramosa* (Figure 1B). Germination stimulation activity of these phenylalkyl ITCs was increased with an increase of the alkyl chain length, and reached a maximum with 4-phenylbutyl ITC and 5-phenylpentyl ITC, being as active as GR24, and dropped dramatically with 6-phenylhexyl ITC (Figure 2B). The germination stimulation activity of 3-phenylpropyl, 4-phenylbutyl, and 5-phenylpentyl ITCs were compared with GR24 at $\leq 10^{-9}$ M, and 5-phenylbutyl ITC appeared to be the most active eliciting about 20% germination even at 10^{-15} M (Figure 3).

In *P. aegyptiaca*, structural requirements of ITCs for germination stimulation were similar to those in *P. ramosa* but not the same (Figure 4A). For example, hexyl ITC induced less than 10% germination of *P. ramosa* (Figure 2A) but elicited about 40% germination of *P. aegyptiaca* (Figure 4A). In contrast, dodecyl ITC induced about 70% germination of *P. ramosa* (Figure 2A) but was almost inactive to *P. aegyptiaca* (Figure 4A). Although benzyl ITC was as active as 2-phenylethyl ITC to *P. ramosa* (Figure 2A), it was less active on *P. aegyptiaca* (Figure 4A). Among phenylalkyl ITCs (Figure 1B), 6-phenylhexyl ITC was inactive and both 4-phenylbutyl ITC and 5-phenylpentyl ITC showed high stimulation activities (Figure 4B) as in the case of *P. ramosa* (Figure 2B).

None of the ITCs tested elicited germination of *Orobanche minor* or *Striga hermonthica* seeds (data not shown).

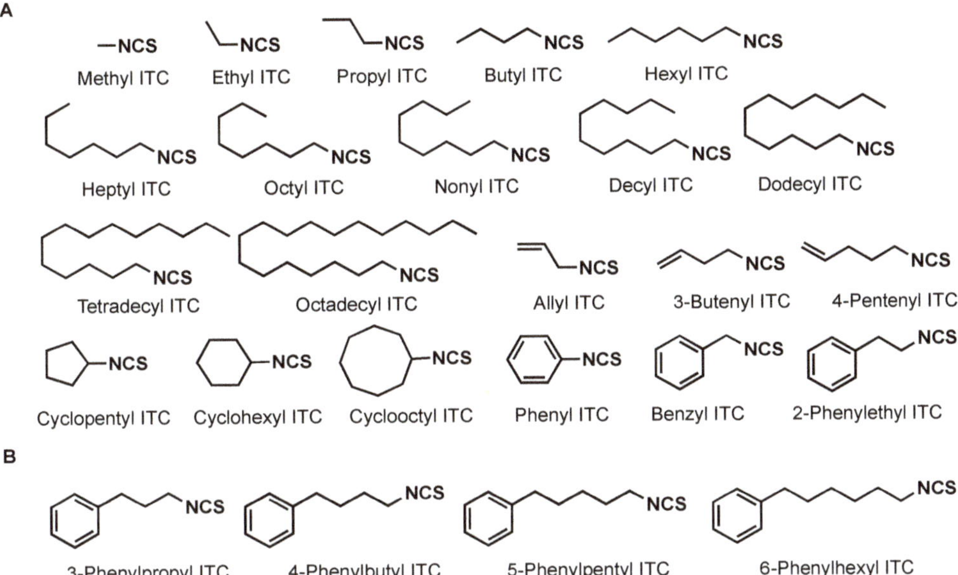

Figure 1. Chemical structures of isothiocyanates used in this study. (**A**) Commercially available products. (**B**) Synthesized in this study.

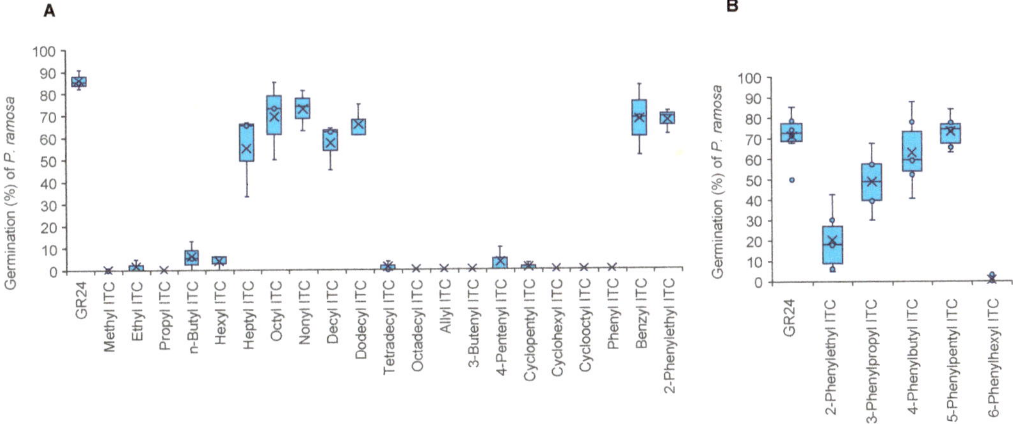

Figure 2. Germination stimulation activities of ITCs on *P. ramosa* seeds. The activities were examined at 10^{-6} M (**A**) and 10^{-8} M (**B**). The box represents the interquartile range, whiskers represent the maximum and minimum values, the middle indicates the median, and the x within the box represents the mean (A, $n = 3$; B, $n = 6$).

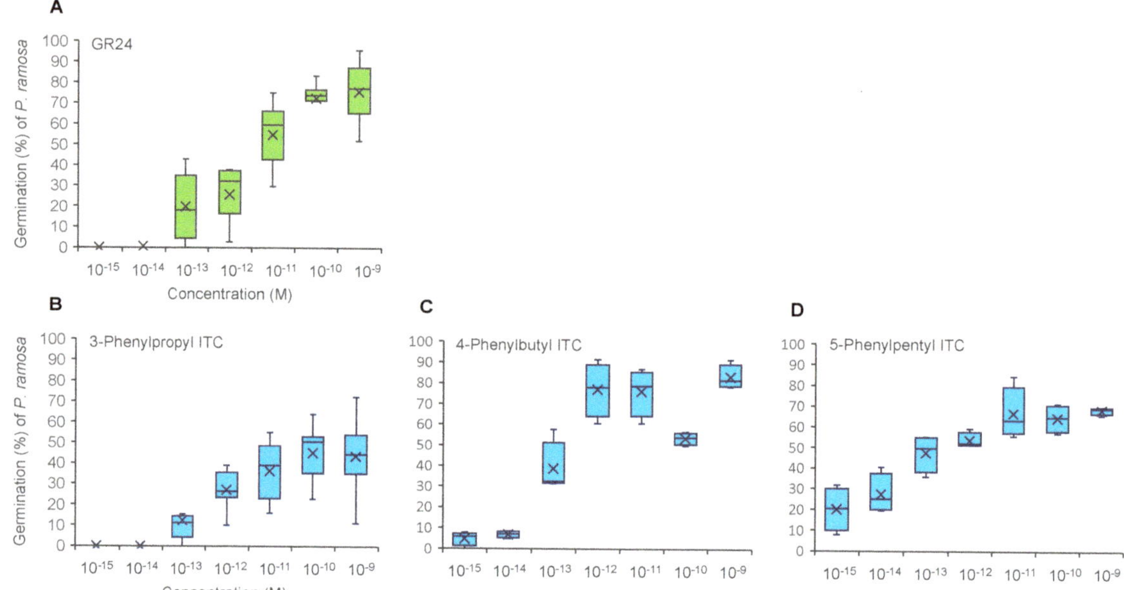

Figure 3. Germination stimulation activities of GR24 (**A**), 3-phenylpropyl ITC (**B**), 4-phenylbutyl ITC (**C**), and 5-phenylpentyl ITC (**D**) on *P. ramosa* at lower concentrations. The box represents the interquartile range, whiskers represent the maximum and minimum values, the middle indicates the median, and the x within the box represents the mean (*n* = 6).

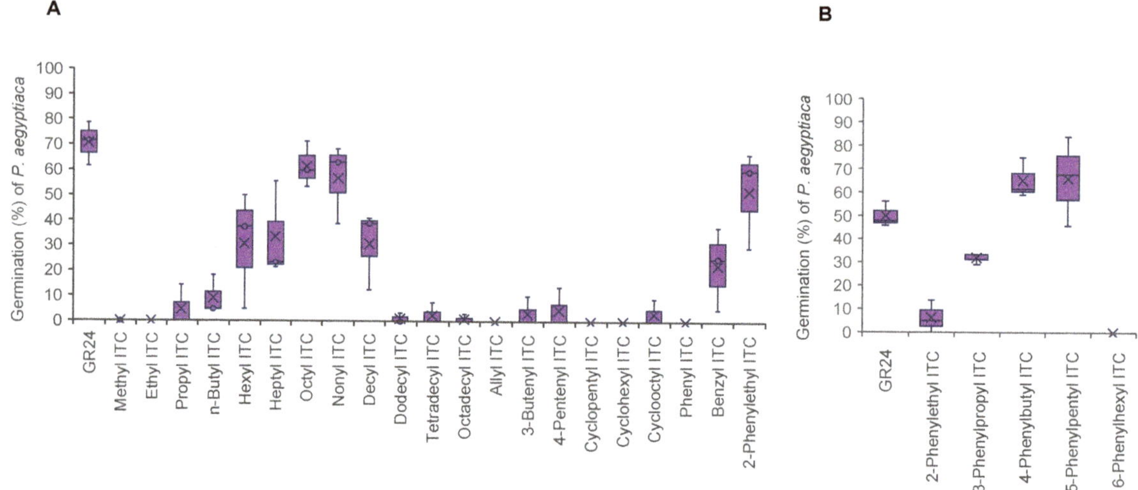

Figure 4. Germination stimulation activities of ITCs on *P. aegyptiaca* seeds. The activities were examined at 10^{-6} M (**A**) and 10^{-8} M (**B**). The box represents the interquartile range, whiskers represent the maximum and minimum values, the middle indicates the median, and the x within the box represents the mean (*n* = 3).

2.2. Host Preference

The seeds of *P. aegyptiaca* collected from mature plants parasitizing different hosts, cabbage, tomato, and chickpea, responded to ITCs (10^{-9} M) with different levels of sensitivity, with those from cabbage hosts most sensitive to ITCs (Figure 5). The seeds from chickpea hosts were moderately sensitive and those from tomato hosts were least sensitive to ITCs. It should be noted that the seeds of *P. aegyptiaca* collected from tomato hosts responded to ITC but not to GR24 at 10^{-9} M.

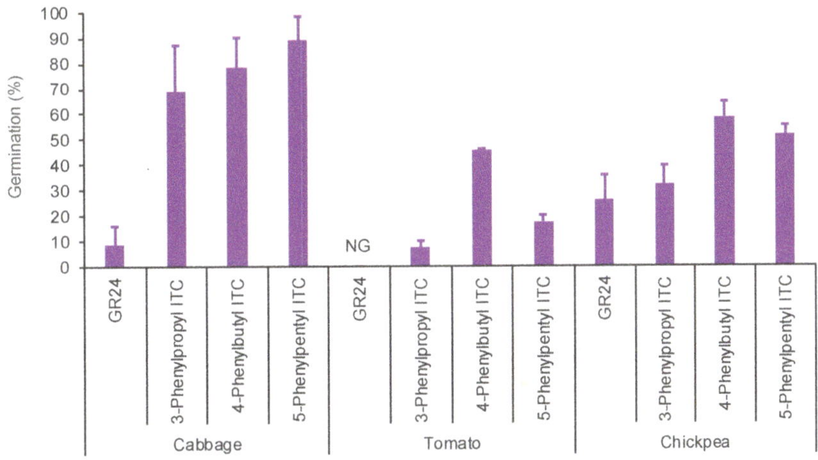

Figure 5. Effects of GR24 and ITCs on germination of *P. aegyptiaca* parasitizing different host plants including cabbage, tomato, and chickpea. Stimulant activities were examined at 10^{-9} M. NG means no germination. Bar means standard errors ($n = 6$).

2.3. Residual Activity and Effects on Germination and Growth of Cabbage

The ITCs 4-Phenylbutyl and 5-phenylpentyl maintained high activity (about 50% that of day 0) after an incubation for 2 weeks in vermiculite (Figure S1). These ITCs did not negatively influence germination or growth of the host plant cabbage (Figure S2).

3. Discussion

Germination stimulation activity of 21 commercially available ITCs indicate that ITCs need an appropriate lipophilicity for high germination stimulation activity and the relatively bulky benzene ring should be separated from the NCS group by at least one carbon (Figure 2). In our previous study, 4-pentenyl ITC was as active as 2-phenylethyl ITC in inducing seed germination of *P. ramosa* [18]. However, in the present study, this ITC was almost inactive (Figure 2). In addition, *P. aegyptiaca* seeds collected from tomato hosts hardly reacted to GR24 in the present study but were rather sensitive to GR24 in the previous study. Such a discrepancy may be due to the different sample application methods in the germination assays; in the previous study, each chemical was applied as an aqueous solution containing 0.1% acetone to the conditioned parasite seeds, whereas in the present study, an aliquot of sample acetone solution was applied to a petri dish lined with a filter paper, water was added to the petri dish after evaporation of the solvent, and then glass fiber disks carrying the conditioned seeds were placed on the filter paper [20]. These differences in the sensitivity to stimulants in the two assay methods, however, may not significantly affect structural requirements in ITCs for germination stimulation.

Alkyl ITCs with an alkyl chain length of C_7 to C_{10} and phenylalkyl ITCs with an alkyl chain length of C_2 to C_5 are highly active germination stimulants for both *P. ramosa* and *P. aegyptiaca*. Therefore, structural requirements of ITCs for *P. ramosa* seed germination are similar to those for *P. aegyptiaca*, but there are clear differences between the two parasitic weed species. For example, at 10^{-6} M hexyl ITC was almost inactive to *P. ramosa* but induced about 20–40% germination in *P. aegyptiaca*. In contrast, dodecyl ITC at 10^{-6} M elicited more than 60% germination in *P. ramosa* but less than 10% germination in *P. aegyptiaca*.

In the case of phenylalkyl ITCs, the benzene ring needed to be separated from the ITC group at least by one carbon atom for high germination stimulation activity. This may suggest that an insertion of an alkyl chain of a proper length between a cycloalkane ring and the ITC group would also enhance the germination stimulation activity (Figures 2A,B and 4A,B). Alternatively, introduction of proper substituents on the benzene ring in the phenylalkyl ITCs may afford highly active stimulants.

One of the promising strategies for mitigating serious damage caused by the root parasitic weeds is to induce suicidal germination in soils [6,19,21]. Sphynolactone-7 (SPL7) was first reported as a femtomolar-range suicide germination stimulant for *S. hermonthica*, which was selected by chemical screening and modified for further activity enhancement [22]. SPL7 possesses methylfuranone but lacks the enol ether bridge, resulting in its enhanced stability. Indeed, in *Striga*-contaminated soil, SPL7 treatment at a concentration of 10^{-10} M a week before planting host maize significantly reduced the emergence of *Striga*. By contrast, GR 24 required 10^{-8} M to achieve similar effects.

In the case of *Striga* seed, ethylene also stimulated germination [6]. Ethylene diffuses widely in the soil, and then ethylene fumigation in *Striga* infested fields has lead to a 90% reduction in viable *Striga* seeds in USA.

Results obtained in in vitro assays do not always support effectiveness in the field, and it is important to examine if the candidate chemicals would work in the fields. GR24, one of the most stable SL analogs, decomposed rather rapidly under field conditions [23,24]. As shown in Figure S1, phenylalkyl ITCs seem to be more stable than SLs in soil, although GR24 was not included in the experiment. In particular, 4-phenylbutyl ITC and 5-phenylpentyl ITC maintained germination stimulation activity about half that of day 0 even after a 14-day incubation.

Methyl ITC generators like dazomet are effective against nematodes and are used widely [25], and in general ITCs are known to exhibit antimicrobial activity [26]. Phenylalkyl ITCs, in particular, benzyl ITC have been reported to be a more active antibiotic than alkyl or alkenyl ITCs including allyl ITC which is known to possess antimicrobial activity against various microorganisms [27–29]. Since benzyl ITC has been shown to be a more active antibiotic than 2-phenylethyl ITC, phenylalkyl ITCs with longer alkyl chains would be weak antimicrobial agents. Therefore, 4-phenylbutyl and 5-phenylpentyl ITCs may be used as suicidal germination inducers for *Phelipanche* spp. but they are only weak antimicrobial agents. There have been several attempts to treat parasitic weed-infested fields with suicidal germination inducers prior to planting host crops [30,31]. In these cases, phytotoxicity of residual germination inducers may cause adverse effects on host crops. The ITCs 4-phenylbutyl and 5-phenylpentyl appeared to be safe for at least one host crop, cabbage (Figure S2).

None of the ITCs examined in this study induced seed germination of *O. minor* or *S. hermonthica* (data not shown). These results demonstrate that ITCs are important germination stimulants for *Phelipanche* spp. which have developed a special seed germination strategy to parasitize *Brassica* spp., non-host plants of AM fungi which exude only small amounts of SLs [15]. Interestingly, there were differences in the sensitivity to ITCs among the seeds of *P. aegyptiaca* parasitizing different host crops (Figure 5); seeds collected from cabbage hosts were highly sensitive to ITCs as compared to the seeds from tomato and chickpea hosts. These results indicate that ITCs would be germination stimulants for *P. aegyptiaca* in the rhizosphere of cabbage. Although tomato and chickpea do not release ITCs and their major germination stimulants are SLs, *P. aegyptiaca* parasitizing these hosts still retains a moderate sensitivity to ITCs. In the case of *P. ramosa*, the seeds collected from different hosts, tobacco and oilseed rape, showed different sensitivities to GR24 [32].

The receptor of SLs in higher plants is the α/β-hydrolase D14, while SL receptor of root parasitic plant is the homolog of D14, namely KARRIKIN INSENSITIVE2/HYPOSENSITIVE

TO LIGHT (KAI2/HTL) [33]. Intriguingly, *S. hermonthica* has eleven *KAI2/HTL* genes and six of them are developed to be highly sensitive to SLs [34]. The expansion of *KAI2/HTL* genes is also observed in *Phelipanche*; five orthologs were identified from both *P. aegyptiaca* [35] and *P. ramosa* [36]. In *P. ramosa*, PrKAI2d3 is likely to be involved in seed germination elicited by both SLs and ITC [36]. PrKAI2d3 has the ability to enzymatically interact with not only SLs but also ITCs, while involvement of other PrKAI2d needs to be clarified. Synthetic SL GR24 was 10,000-fold more active than the ITCs in both germination stimulation and interaction with the receptor PrKI2d3. The high germination stimulation activities of ITCs observed in our experiments suggest that these ITCs with an appropriate lipophilicity would permeate into lipid-rich parasite seeds more easily than do SLs [37]. Further structural optimization of ITCs, for example, introduction of substituent(s) onto the benzene ring of 4-phenylbutyl ITC and 5-phenylpentyl ITC (Figure 1B) may afford more active ITCs with longer residual activities.

Recently, non-SL germination stimulants for *Orobanche minor* have been isolated from *Streptomyces albus* J1074 [38] and tryptophan derivatives have been shown to induce *O. minor* seed germination [39]. Although these compounds required high concentrations to induce germination, structural modifications may enhance their activities. It remains unclear if these compounds induce germination of other root parasitic weeds. It is interesting that each root parasitic plant species appears to have flexibly evolved to become sensitive to specific chemicals in their specific environments.

4. Materials and Methods

4.1. Plant Material

P. ramosa seeds were collected from mature flowering spikes that were parasites of oilseed rape grown at Saint-Martin-de Fraigneau, France. *P. aegyptiaca* seeds were collected in tomato, cabbage, and chickpea crops at Golan Heights, Western Negev, and Western Galilee, respectively, in Israel. *O. minor* was collected from a red clover field at Utsunomiya in Japan and *S. hermonthica* from a maize field near Wad Medani in Sudan. Seeds of cabbage were obtained from a local supplier.

4.2. Chemicals

A total of 21 commercially available ITCs were obtained from Tokyo Chemical Industry Co. Ltd. (Tokyo, Japan). GR24 was kindly supplied by Prof. Kohki Akiyama (Osaka Prefecture University, Osaka, Japan). The other analytical grade chemicals were obtained from Kanto Chemical Co. Ltd. (Tokyo, Japan).

4.3. Synthesis

The ITCs 3-Phenylpropyl-, 4-phenylbutyl-, 5-phenylpentyl- and 6-phenylhexyl were prepared through the decomposition of dithiocarbamic acid salts generated in situ by the treatment of the corresponding amines with carbon disulfide and triethylamine [40,41] (Figure S3).

4.4. Germination Assay

The germination assay was conducted as reported previously [20]. ITCs were dissolved and diluted by using acetone. An aliquot (<10 µL) of the respective ITC solution was added to each petri dish (i.d. 5 cm) lined with a filter paper. The solvent was allowed to evaporate before conditioned seeds were placed on the filter paper and treated with distilled water (0.65 mL). The seeds of *P. ramosa*, *P. aegyptiaca*, *O. minor*, and *S. hermonthica* seeds were incubated at their optimal temperature, 23, 25, 23, and 30 °C, respectively.

4.5. Residual Activity

ITCs dissolved in water (100 mL) were applied onto pots (i.d. 10 cm, 10 cm deep) filled with vermiculite (300 g). Pots were incubated for 2, 7, and 14 days without any host plants under 150 μm m^{-2} s^{-1} (16 h light and 5 h dark) at room temperature. After incubation, the pot was washed with 200 mL of water, the drain water was collected, and extracted with 100 mL of ethyl acetate three times. The ethyl acetate solutions were combined, dried over anhydrous Na_2SO_4, and concentrated in vacuo. The ethyl acetate extracts were examined for germination stimulation of *P. aegyptiaca* as in Section 4.4.

4.6. Effects of ITCs on Germination and Growth of Cabbage

Cabbage seeds were incubated with 4-phenylbutyl and 5-phenylpentyl ITC (10^{-8}–10^{-6} M) at 20 °C for 5 days and the germination rate and shoot and root lengths were determined.

4.7. Statistical Analysis

All experimental results were subjected to ANOVA utilizing JMP software, version 5.0 (SAS Institute Inc., Cary, NC, USA).

5. Conclusions

We found highly active germination stimulant ITCs to *Phelipanche* spp., which induce germination as low as at 1 femtomolar. They seem to be relatively stable in soils and do not negatively affect germination or growth of the cabbage host plant. These results suggest that these ITCs would be good lead compounds for suicidal germination inducers to control *Phelipanche* spp. in the field. The seeds of *P. aegyptiaca* collected from mature plants parasitizing cabbage, tomato, and chickpea were more sensitive to these ITCs than the synthetic SL GR24. Hence, ITCs have the potential to be used as inducers of suicidal germination of *Phelipanche* spp.

Supplementary Materials: The following supporting information can be downloaded at: https://www.mdpi.com/article/10.3390/plants11050606/s1, Figure S1: Residual activity of phenylalkyl ITCs on *P. aegyptiaca* germination, Figure S2: Effects of ITCs on germination and growth of host plant cabbage, Figure S3: Synthetic schemes of ITCs.

Author Contributions: K.Y. (Kaori Yoneyama) and H.M. conducted the experiment; K.Y. (Kaori Yoneyama), H.M., R.O., H.N. (Hidemitsu Nakamura), H.N. (Hisashi Nishiwaki), S.Y., X.X. and K.Y. (Koichi Yoneyama) designed the experiments; K.Y. (Kaori Yoneyama) and H.M. analyzed the data; K.Y. (Kaori Yoneyama), H.M., and K.Y. (Koichi Yoneyama) wrote the manuscript. All authors have read and agreed to the published version of the manuscript.

Funding: This work was supported by the Japan Society for the Promotion of Science (KAKENHI 15J40043,16K18560, and 21H02125), and the Japan Science and Technology Agency, Precursory Research for Embryonic Science and Technology (PRESTO, JPMJPR17QA).

Institutional Review Board Statement: Not applicable.

Informed Consent Statement: Not applicable.

Data Availability Statement: Data is contained within the article and supplementary material.

Acknowledgments: We appreciate Amit Wallach and Yakkov Goldwasser (The Hebrew University of Jerusalem), Jean-Bernard Pouvreau (University of Nantes), and A.G.T. Babiker (Sudan University of Science and Technology) for supplying seeds of *P. aegyptiaca*, *P. ramosa*, and *S. hermonthica*, respectively.

Conflicts of Interest: The authors declare no conflict of interest.

References

1. Parker, C.; Riches, C.R. *Parasitic Weeds of the World: Biology and Control*; CAB International: Wallingford, UK, 1993; p. 332.
2. Musselman, L.J. The biology of *Striga*, *Orobanche*, and other root-parasitic weeds. *Annu. Rev. Phytopathol.* **1980**, *18*, 463–489. [CrossRef]
3. Sauerborn, J. The economic importance of the phytoparasites *Orobanche* and *Striga*. In Proceedings of the 5th International Symposium of Parasitic Weeds, Nairobi, Kenya, 24–30 June 1991; pp. 137–143.
4. Joel, D.M. The long-term approach to parasitic weeds control: Manipulation of specific developmental mechanisms of the parasite. *Crop Prot.* **2000**, *19*, 753–758. [CrossRef]
5. Xie, X.; Yoneyama, K.; Yoneyama, K. The strigolactone story. *Annu. Rev. Phytopathol.* **2010**, *48*, 93–117. [CrossRef] [PubMed]
6. Eplee, R.E. Ethylene: A witchweed seed germination stimulant. *Weed Sci.* **1975**, *23*, 433–436. [CrossRef]

7. Al-Babili, S.; Bouwmeester, H.J. Strigolactones, a novel carotenoid-derived plant hormone. *Annu. Rev. Plant Biol.* **2015**, *66*, 161–186. [CrossRef] [PubMed]
8. Cook, C.E.; Whichard, L.P.; Turner, B.; Wall, M.E.; Egley, G.H. Germination of witchweed (*Striga lutea* Lour.): Isolation and properties of a potent stimulant. *Science* **1966**, *154*, 1189–1190. [CrossRef]
9. Yoneyama, K. Recent progress in the chemistry and biochemistry of strigolactones. *J. Pestic. Sci.* **2020**, *45*, 45–53. [CrossRef]
10. Yoneyama, K.; Brewer, P.B. Strigolactones, how are they synthesized to regulate plant growth and development? *Curr. Opin. Plant Biol.* **2021**, *63*, 102072. [CrossRef]
11. Akiyama, K.; Matsuzaki, K.; Hayashi, H. Plant sesquiterpenes induce hyphal branching in arbuscular mycorrhizal fungi. *Nature* **2005**, *435*, 824–827. [CrossRef]
12. Besserer, A.; Puech-Pagès, V.; Kiefer, P.; Gomez-Roldan, V.; Jauneau, A.; Roy, S.; Portais, J.C.; Roux, C.; Bécard, G.; Séjalon-Delmas, N. Strigolactones stimulate arbuscular mycorrhizal fungi by activating mitochondria. *PLoS Biol.* **2006**, *4*, e226. [CrossRef]
13. Soto, M.J.; Fernández-Aparicio, M.; Castellanos-Morales, V.; Garciía-Garrido, J.M.; Ocampo, J.A.; Delgado, M.J.; Vierheilig, H. First indications for the involvement of strigolactones on nodule formation in alfalfa (*Medicago sativa*). *Soil Biol. Biochem.* **2010**, *42*, 383–385. [CrossRef]
14. Foo, E.; Davies, N.W. Strigolactones promote nodulation in pea. *Planta* **2011**, *234*, 1073–1081. [CrossRef]
15. Yoneyama, K.; Xie, X.; Sekimoto, H.; Takeuchi, Y.; Ogasawara, S.; Akiyama, K.; Hayashi, H.; Yoneyama, K. Strigolactones, host recognition signals for root parasitic plants and arbuscular mycorrhizal fungi, from Fabaceae plants. *New Phytol.* **2008**, *179*, 484–494. [CrossRef] [PubMed]
16. Kalia, V.C.; Gong, C.; Patel, S.K.S.; Lee, J.K. Regulation of plant mineral nutrition by signal molecules. *Microorganisms* **2021**, *9*, 774. [CrossRef] [PubMed]
17. Smith, S.E.; Read, D.J. *Mycorrhizal Symbiosis*, 3rd ed.; Academic Press: Cambridge, UK, 2008; p. 787.
18. Auger, B.; Pouvreau, J.-B.; Pouponneau, K.; Yoneyama, K.; Montiel, G.; Le Bizec, B.; Yoneyama, K.; Delavault, P.; Delourme, R.; Simier, P. Germination stimulants of *Phelipanche ramosa* in the rhizosphere of *Brassica napus* are derived from the glucosinolate pathway. *Mol. Plant-Microbe Interact.* **2012**, *25*, 993–1004. [CrossRef]
19. Dubey, S.; Guignard, F.; Pellaud, S.; Pedrazzetti, M.; van der Schuren, A.; Gaume, A.; Schnee, S.; Gindro, K.; Dubey, O. Isothiocyanate derivatives of glucosinolates as efficient natural fungicides. *PhytoFrontiers* **2021**, *1*, 40–50. [CrossRef]
20. Yoneyama, K.; Takeuchi, Y.; Ogasawara, M.; Konnai, M.; Sugimoto, Y.; Sassa, T. Cotylenins and fusicoccins stimulate seed germination of *Striga hermonthica* (Del.) Benth and *Orobanche minor* Smith. *J. Agric. Food Chem.* **1998**, *46*, 1583–1586. [CrossRef]
21. Kountche, B.A.; Jamil, M.; Yonli, D.; Nikiema, M.P.; Blanco-Ania, D.; Asami, T.; Zwanenburg, B.; Al-Babili, S. Suicidal germination as a control strategy for *Striga hermonthica* (Benth.) in smallholder farms of sub-Saharan Africa. *Plants People Planet* **2019**, *1*, 107–118. [CrossRef]
22. Uraguchi, D.; Kuwata, K.; Hijikata, Y.; Yamaguchi, R.; Imaizumi, H.; AM, S.; Rakers, C.; Mori, N.; Akiyama, K.; Irie, S.; et al. A femtomolar-range suicide germination stimulant for the parasitic plant *Striga hermonthica*. *Science* **2018**, *362*, 1301–1305. [CrossRef]
23. Babiker, A.G.T.; Ibrahim, N.E.; Edwards, W.G. Persistence of GR7 and *Striga* germination stimulant(s) from *Euphorbia aegyptiaca* Boiss. in soils and in solutions. *Weed Res.* **1988**, *28*, 1–6. [CrossRef]
24. Halouzka, R.; Tarkowski, P.; Zwanenburg, B.; Cavar Zeljkovic, S. Stability of strigolactone analog GR24 toward nucleophiles. *Pest Manag. Sci.* **2018**, *74*, 896–904. [CrossRef] [PubMed]
25. Zasada, I.A.; Halbrendt, J.M.; Kokalis-Burelle, N.; LaMondia, J.; McKenry, M.V.; Noling, J.W. Managing nematodes without methyl bromide. *Annu. Rev. Phytopathol.* **2010**, *48*, 311–328. [CrossRef] [PubMed]
26. Romeo, L.; Iori, R.; Rollin, P.; Bramanti, P.; Mazzon, E. Isothiocyanates: An overview of their antimicrobial activity against human infections. *Molecules* **2018**, *23*, 624. [CrossRef]
27. Dias, C.; Aires, A.; Saavedra, M.J. Antimicrobial activity of isothiocyanates from cruciferous plants against methicillin-resistant *Staphylococcus aureus* (MRSA). *Int. J. Mol. Sci.* **2014**, *15*, 19552–19561. [CrossRef] [PubMed]
28. Dufour, V.; Alazzam, B.; Ermel, G.; Thepaut, M.; Rossero, A.; Tresse, O.; Baysse, C. Antimicrobial activities of isothiocyanates against *Campylobacter jejuni* isolates. *Front. Cell. Infect. Microbiol.* **2012**, *2*, 53. [CrossRef] [PubMed]
29. Li, D.; Shu, Y.; Li, P.; Zhang, W.; Ni, H.; Cao, Y. Synthesis and structure–activity relationships of aliphatic isothiocyanate analogs as antibiotic agents. *Med. Chem. Res.* **2013**, *22*, 3119–3125. [CrossRef]
30. Zwanenburg, B.; Mwakaboko, A.S.; Kannan, C. Suicidal germination for parasitic weed control. *Pest Manag. Sci.* **2016**, *72*, 2016–2025. [CrossRef]
31. Samejima, H.; Babiker, A.G.; Takikawa, H.; Sasaki, M.; Sugimoto, Y. Practicality of the suicidal germination approach for controlling *Striga hermonthica*. *Pest Manag. Sci.* **2016**, *72*, 2035–2042. [CrossRef]
32. Gibot-Leclerc, S.; Connault, M.; Perronne, R.; Dessaint, F. Differences in seed germination response of two populations of *Phelipanche ramosa* (L.) Pomel to a set of GR24 concentrations and durations of stimulation. *Seed Sci. Res.* **2021**, *31*, 243–248. [CrossRef]
33. Waters, M.T.; Gutjahr, C.; Bennett, T.; Nelson, D.C. Strigolactone signaling and evolution. *Annu. Rev. Plant Biol.* **2017**, *68*, 291–322. [CrossRef]
34. Toh, S.; Holbrook-Smith, D.; Stogios, P.J.; Onopriyenko, O.; Lumba, S.; Tsuchiya, Y.; Savchenko, A.; McCourt, P. Structure-function analysis identifies highly sensitive strigolactone receptors in *Striga*. *Science* **2015**, *350*, 203–207. [CrossRef] [PubMed]

35. Conn, C.E.; Bythell-Douglas, R.; Neumann, D.; Yoshida, S.; Whittington, B.; Westwood, J.H.; Shirasu, K.; Bond, C.S.; Dyer, K.A.; Nelson, D.C. Convergent evolution of strigolactone perception enabled host detection in parasitic plants. *Science* **2015**, *349*, 540–543. [CrossRef]
36. de Saint Germain, A.; Jacobs, A.; Brun, G.; Pouvreau, J.-B.; Braem, L.; Cornu, D.; Clavé, G.; Baudu, E.; Steinmetz, V.; Servajean, V.; et al. A *Phelipanche ramosa* KAI2 protein perceives enzymatically strigolactones and isothiocyanates. *Plant Commun.* **2021**, *2*, 100166. [CrossRef] [PubMed]
37. Joel, D.M.; Bar, H. The Seed and the Seedling. In *Parasitic Orobanchaceae: Parasitic Mechanisms and Control Strategies*; Joel, D.M., Gressel, J., Musselman, L.J., Eds.; Springer: Berlin/Heidelberg, Germany, 2013; pp. 147–165.
38. Okazawa, A.; Samejima, H.; Kitani, S.; Sugimoto, Y.; Ohta, D. Germination stimulatory activity of bacterial butenolide hormones from *Streptomyces albus* J1074 on seeds of the root parasitic weed *Orobanche minor*. *J. Pestic. Sci.* **2021**, *46*, 242–247. [CrossRef] [PubMed]
39. Kuruma, M.; Suzuki, T.; Seto, Y. Tryptophan derivatives regulate the seed germination and radicle growth of a root parasitic plant, *Orobanche minor*. *Bioorg. Med. Chem. Lett.* **2021**, *43*, 128085. [CrossRef]
40. Morse, M.A.; Eklind, K.I.; Hecht, S.S.; Jordan, K.G.; Choi, C.-I.; Desai, D.H.; Amin, S.G.; Chung, F.-L. Structure-activity relationships for inhibition of 4-(methylnitrosamino)-1-(3-pyridyl)-1-butanone lung tumorigenesis by arylalkyl isothiocyanates in A/J mice. *Cancer Res.* **1991**, *51*, 1846–1850.
41. Wong, R.; Dolman, S.J. Isothiocyanates from tosyl chloride mediated decomposition of in situ generated dithiocarbamic acid salts. *J. Org. Chem.* **2007**, *72*, 3969–3971. [CrossRef]

Article

A New Formulation for Strigolactone Suicidal Germination Agents, towards Successful *Striga* Management

Muhammad Jamil [1], Jian You Wang [1], Djibril Yonli [2], Rohit H. Patil [3], Mohammed Riyazaddin [4], Prakash Gangashetty [4], Lamis Berqdar [1], Guan-Ting Erica Chen [1,5], Hamidou Traore [2], Ouedraogo Margueritte [2], Binne Zwanenburg [6], Satish Ekanath Bhoge [3] and Salim Al-Babili [1,5,*]

[1] The BioActives Lab, Center for Desert Agriculture, King Abdullah University of Science and Technology, Thuwal 23955-6900, Saudi Arabia; muhammad.jamil@kaust.edu.sa (M.J.); jianyou.wang@kaust.edu.sa (J.Y.W.); lamis.berqdar@kaust.edu.sa (L.B.); guanting.chen@kaust.edu.sa (G.-T.E.C.)

[2] Institut de l'Environnement et de Recherches Agricoles (INERA), Ouagadougou 04 BP 8645, Burkina Faso; d.yonli313@gmail.com (D.Y.); hamitraore8@yahoo.com (H.T.); margoued616@gmail.com (O.M.)

[3] UPL House, Express Highway, Bandra-East, Mumbai 400 051, Maharashtra, India; rohit.patil2@upl-ltd.com (R.H.P.); bhogese@upl-ltd.com (S.E.B.)

[4] International Crops Research Institute for the Semi-Arid Tropics (ICRISAT), Niamey BP 12404, Niger; m.riyazaddin@cgiar.org (M.R.); p.gangashetty@cgiar.org (P.G.)

[5] Plant Science Program, Biological and Environmental Science and Engineering Division, King Abdullah University of Science and Technology (KAUST), Thuwal 23955-6900, Saudi Arabia

[6] Institute for Molecules and Materials, Radboud University, 6525 AJ Nijmegen, The Netherlands; b.zwanenburg@science.ru.nl

* Correspondence: salim.babili@kaust.edu.sa

Citation: Jamil, M.; Wang, J.Y.; Yonli, D.; Patil, R.H.; Riyazaddin, M.; Gangashetty, P.; Berqdar, L.; Chen, G.-T.E.; Traore, H.; Margueritte, O.; et al. A New Formulation for Strigolactone Suicidal Germination Agents, towards Successful *Striga* Management. *Plants* **2022**, *11*, 808. https://doi.org/10.3390/plants 11060808

Academic Editors: Evgenia Dor and Yaakov Goldwasser

Received: 13 February 2022
Accepted: 15 March 2022
Published: 18 March 2022

Publisher's Note: MDPI stays neutral with regard to jurisdictional claims in published maps and institutional affiliations.

Abstract: *Striga hermonthica*, a member of the *Orobanchaceae* family, is an obligate root parasite of staple cereal crops, which poses a tremendous threat to food security, contributing to malnutrition and poverty in many African countries. Depleting *Striga* seed reservoirs from infested soils is one of the crucial approaches to minimize subterranean damage to crops. The dependency of *Striga* germination on the host-released strigolactones (SLs) has prompted the development of the "Suicidal Germination" strategy to reduce the accumulated seed bank of *Striga*. The success of aforementioned strategy depends not only on the activity of the applied SL analogs, but also requires suitable application protocol with simple, efficient, and handy formulation for rain-fed African agriculture. Here, we developed a new formulation "Emulsifiable Concentration (EC)" for the two previously field-assessed SL analogs Methyl phenlactonoate 3 (MP3) and Nijmegen-1. The new EC formulation was evaluated for biological activities under lab, greenhouse, mini-field, and field conditions in comparison to the previously used Atlas G-1086 formulation. The EC formulation of SL analogs showed better activities on *Striga* germination with lower EC_{50} and high stability under Lab conditions. Moreover, EC formulated SL analogs at 1.0 µM concentrations reduced 89–99% *Striga* emergence in greenhouse. The two EC formulated SL analogs showed also a considerable reduction in *Striga* emergence in mini-field and field experiments. In conclusion, we have successfully developed a desired formulation for applying SL analogs as suicidal agents for large-scale field application. The encouraging results presented in this study pave the way for integrating the suicidal germination approach in sustainable *Striga* management strategies for African agriculture.

Keywords: *Striga hermonthica*; seedbank; suicidal germination; strigolactone analogs; witch weeds; methyl phenlactonoate

1. Introduction

Striga hermonthica, an obligate root-parasitic weed, is one of the major biotic constraints to cereal production in sub-Saharan Africa [1–4]. After spreading in about 32 African countries, *Striga* has infested about 50 million hectares of arable land in Africa [5,6]. The crop yield losses due to *Striga* infestation can vary from 40 to 100%, causing annual losses of around US $10 billion and threatening the life and food security of 300 million African people [7–11]. Developing suitable *Striga* control strategies is crucial for not only reducing the extent of damage but also retaining further spread into the non-contaminant fields [12]. However, the management of *Striga* is challenging due to the production of ~0.1 million seeds per plant [13], up to 20 years of seed longevity [14], complex life cycle [3], underground damage [15], and host dependency of *Striga* seed germination [16].

The germination of *Striga* seed requires favorable hot and humid conditions [17] and, most importantly, the perception of host-released chemical signals, such as strigolactones (SLs) [18–20]. The germinated *Striga* seeds should establish a connection to the root system of the host to survive due to very limited food reserve in tiny seeds for a short period of time [3,21]. This essential step of *Striga* life cycle leads to a basis for a promising control strategy, known as "Suicidal Germination" [22,23]. The germination of *Striga* seeds can be triggered in the bare soil by direct application of synthetic germination stimulants [24,25]. This germination in the host's absence is lethal for *Striga*, which can be exploited as a tool to eliminate the accumulated *Striga* seed bank [22,23,26]. Although this concept has been proposed in a number of previous studies [22,23,26,27], but the availability of simple, easy-to-synthesize, and affordable synthetic SL analogs with suitable formulation and application under natural, hot, and rain-fed field conditions is still arguable [28]. Indeed, the selection of suitable germination stimulants, application protocol, and appropriate formulation for field application remain as challenging barriers to the success of this technology.

Several SL analogs and mimics were developed over the last three decades, but the problem of their efficacy, stability, and synthesis remains unsolved [29–31]. To identify suitable suicidal germination stimulants, a few SL analogs have been screened during the past 10 years based on the bioactivity of *Striga* germination under different conditions [25,32–34]. Indeed, two potent SL analogs, namely, MP3 and Nijmegen-1, were selected to be tested for suicidal germination activity under field conditions [23,32]. Another crucial step was to devise an appropriate protocol for successful field application. Keeping in mind that scarcity of water, type of soil, frequency and concentration of application, a rain-fed based suicidal application protocol has been proposed recently [23]. In addition, the development of active and viable formulation, suitable for harsh African agricultural conditions, is very crucial for the success of suicidal technology. Although the use of SL analogs formulated in Atlas G-1086 (AG), by Coroda Crop Care, The Netherlands, has been reported in previous studies, this emulsifier is expensive and time-consuming for preparation. Moreover, the solubility problems of active ingredients in the AG formulation and side effects on the host crop have been observed. Alternately, another formulation "Emulsifiable Concentrate (EC)" was developed by the UPL, India. The present study is carried out to test and evaluate the efficacy of this newly developed EC formulation of two very simple and efficient SL analogs under laboratory, greenhouse, and field conditions. We also compared the old AG formulations of both SL analogs on the suicidal germination activity with this new formulation. Besides, this is the first SL analogs formulation as suicidal agents, synthesized on a large-scale for real field application. This EC formulation is user-friendly, handy, water-soluble, highly compatible with SL analogs, and stable at room temperature (>2 years). To this end, our results will lead to developing a package of suicidal technology to combat *Striga* in Africa.

2. Results

2.1. Striga Seed Germination in Response to EC and AG Formulations of SL Analogs under Lab Conditions

Structure of SL analogs and packing of Emulsifiable Concentrate (EC) formulation of MP3 27EC and Nijmegen 34EC are shown in Figure 1. The scheme of the experiment conducted for *Striga* seed germination bioassays is depicted in Figure 2A. The EC formulation of SL analogs MP3 27EC and Nijmegen 34EC induced about 56–59% *Striga* seed germination at a concentration of 1.0 µM, which was about 3–13% higher than AG formulation (49–57%) (Figure 2B). We also compared the activity of the two EC formulated analogs MP3

27EC and Nijmegen 34EC and observed both analogs at 1, 0.1, and 0.01 µM had similar activity, while Nijmegen 34EC at lower 0.001 µM showed a better activity and exhibited the lower EC_{50} value (0.008 vs. 0.036 µM) (Figure 2C). Importantly, EC formulation effectively reduced the value of EC_{50} in comparison to AG formulation (Figure 2C). Moreover, both EC formulated SL analogs induced seed germination of various *Striga* batches collected from Kenya, Niger, and Burkina Faso. Seed collected from Kenya showed about 9–15% *Striga* seed germination, followed by Burkina Faso batch with 11–13%, and Niger batch with 8–11% germination (Figure 3). The *Striga* seeds collected from Sudan appeared very active, showing maximum germination (~60%) as compared to seed populations collected from Burkina Faso and Niger (8–11%). However, these outcomes revealed that EC formulation of both SL analogs are still able to induce germination of various ecotypes of *Striga* seeds throughout African countries, depending upon seed viability and dormancy.

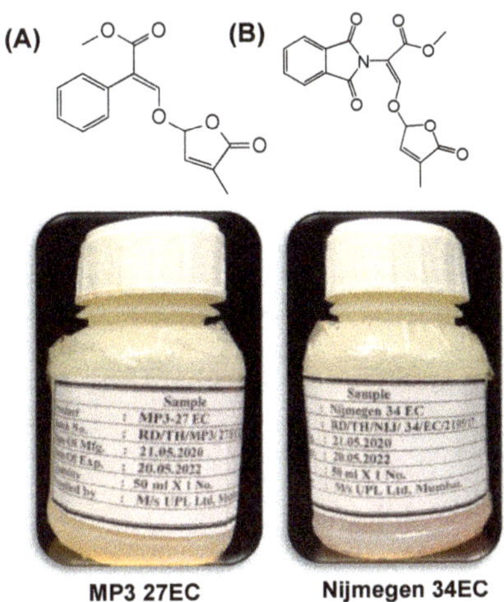

Figure 1. Structure of selected SL analogs and packaging of Emulsifiable Concentrate (EC) formulation of strigolactone analogs, developed by UPL, India (**A**) MP3 27EC and (**B**) Nijmegen 34EC, prepared by UPL, India.

2.2. Stability of EC and AG Formulations of Strigolactone Analogs

Fresh preparations of 1.0 µM SL analogs MP3 and Nijmegen with both formulations demonstrated about 52–64% *Striga* germination. EC formulation exhibited 60–64% germination when compared to AG formulation 52–57% (Figure 4). MP3 in EC formulation remained active even at 12 weeks after application, showing up to 51% *Striga* seed germination; while the AG formulation of MP3 only had a 29% activity on *Striga* germination on week 12. Surprisingly, Nijmegen in AG formulation completely lost its activity in week 6 (~30% *Striga* germination on Week 4); whereas EC formulation of Nijmegen was much more stable and remained active up to week 6 with ~29% *Striga* germination (Figure 4).

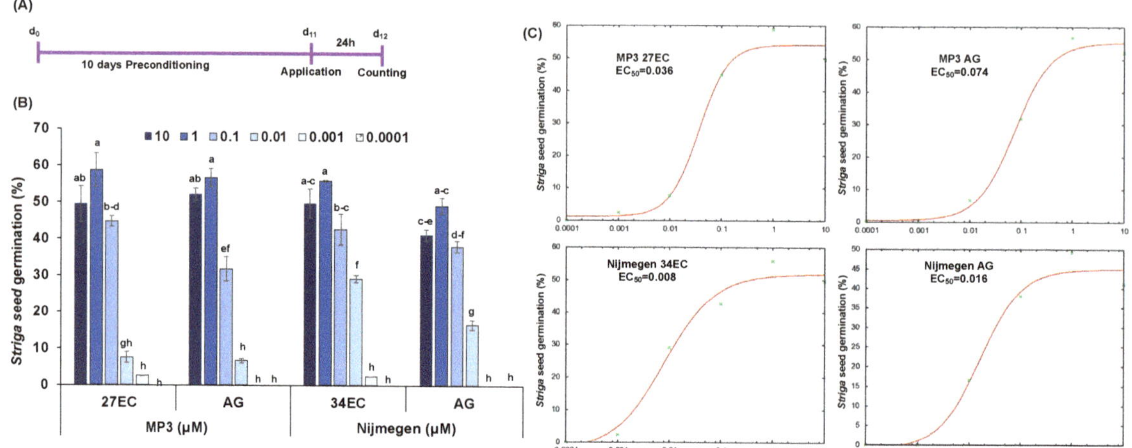

Figure 2. Comparison of EC formulated SL analogs MP3 27EC and Nijmegen 34EC with AG formulated SL analogs for *Striga* seed germination. (**A**) The scheme of the experiment conducted for *Striga* seed germination bioassays. (**B**) Effect of various concentrations of both formulations (AG and EC) of the two SL analogs on *Striga* seed germination. The *Striga* seeds collected from a sorghum infested field during 2020 in Sudan were treated with the two formulations of SL analogs. (**C**) EC_{50} of *Striga* seed germination in response to the various concentrations of the two SL analogs in different formulations. Data are means $\pm$ SE (n = 3), treatments with various letters differ significantly according to one-way analysis of variance (ANOVA) and Tukey's post hoc test ($p < 0.05$).

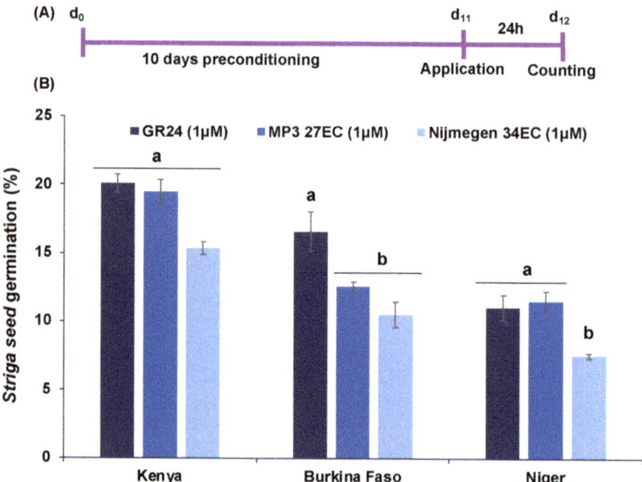

Figure 3. Effect of EC formulated SL analogs (MP3 27EC and Nijmegen 34EC) on *Striga* seed germination collected from a maize field in Kenya, pearl millet fields in Burkina Faso, and Niger. (**A**) Scheme of the experiment conducted for *Striga* seed germination bioassays. (**B**) *Striga* seed germination of various ecotype in response to MP3 27EC and Nijmegen 34EC. Data are means $\pm$ SE (n = 7). For each SL analog, treatments with various letters differ significantly according to one-way analysis of variance (ANOVA) and Tukey's post hoc test ($p < 0.05$).

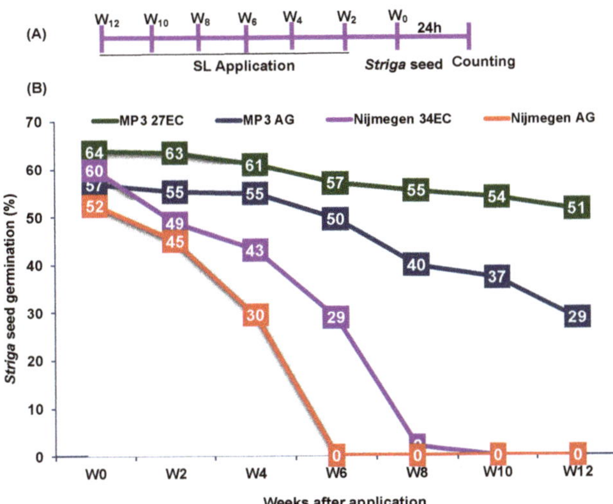

Figure 4. Stability of EC and AG formulation of the two strigolactone analogs. (**A**) Scheme of the experiment conducted biweekly to test the stability of SL analogs. (**B**) *Striga* seeds germination in response to two formulations of MP3 and Nijmegen at two-week intervals. Each SL analog (10 mL) was applied in a Petri plate on 5 filter papers and incubated at 30 °C for two-week intervals up to 12 weeks. The *Striga* seeds collected from a sorghum field in Sudan were evaluated for germination in each Petri plate. Data are means ± SE (*n* = 5).

2.3. Effect of Various Formulations of Strigolactone Analogs on Striga Emergence in Pots under Greenhouse Conditions

Both formulations of the two SL analogs (MP3 and Nijmegen) were further evaluated by applying at 1.0 µM concentration to *Striga* infested pot under greenhouse conditions (Figure 5A,B). Intriguingly, EC formulation of both analogs showed around 89–99% decrease of *Striga* emergence, which was about 3–5% higher than the reduction detected with the AG formulation (85–96%) (Figure 5C). However, no clear difference was obtained between the two formulated Nijmegen groups. Among the two EC formulated analogs, MP3 27EC had a larger decline in *Striga* emergence (99%) than Nijmegen 34EC (89%). In addition, a higher reduction in *Striga* emergence led to a better growth of the host plant, indicated by an increase of 61–99% in plant height of the host crop (Figure 5D).

2.4. Effect of Various Formulations of Strigolactone Analogs on Striga Infection under Mini-Field Conditions

To fulfill the practical purpose of developing EC formulation for alleviating *Striga*, the efficacy of the two SL analogs MP3 27EC and Nijmegen 34EC was tested under mini-field conditions (Figure 6A–C) in INERA, Burkina Faso and ICRISAT, Niger. At INERA, although a huge variation among replications resulted in a non-significant impact on *Striga* germination (Figure 6D), apparently, we observed more *Striga* germination upon the application of MP3 27EC and Nijmegen 34EC (at 1.0 µM) compared to AG formulation. EC formulation of both SLs showed about 15–31% *Striga* germination and the Nijmegen 34EC appeared more active than MP3 27EC (Figure 6D). In addition to *Striga* germination, EC formulation of both SL analogs showed 9–23% reduction in *Striga* emergence, whereas we did not observe any reduction in *Striga* emergence after AG formulation treatment (Figure 6E). Similarly, we noticed a huge variation among replicated mini-boxes from ICRISAT, Niger, making the results non-significant. However, we still observed 35–53% reduction in *Striga* emergence by EC formulation of MP3 27EC and Nijmegen 34EC which was better than the AG formulation (16–21%). Additionally, Nijmegen 34EC showed better activity than MP3 27EC (Figure 7).

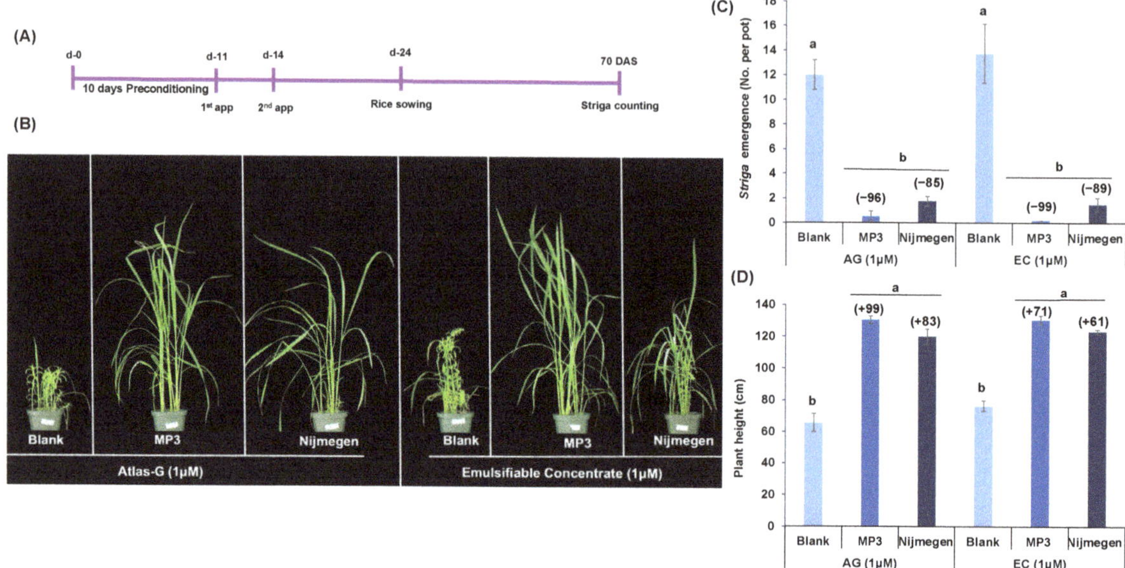

Figure 5. Effect of various formulations of strigolactone analogs on *Striga* emergence in pots under greenhouse conditions. (**A**) Scheme of the experiment conducted for *Striga* emergence in pots under greenhouse conditions. (**B**) View of the *Striga* infected pots. Each pot was filled with the soil infested with *Striga* seeds, collected from a sorghum field in Sudan. Both formulations of the two SL analogs (at 1.0 µM) were applied for two times in *Striga* infested pots. *Striga* emergence was counted at 70 days after sowing of rice. (**C**) Values of each bar showed average emergence of *Striga* per plot. (**D**) Average plant height of rice host plant measured at 70 DAS. Data are means ± SE ($n = 4$). For each SL analog, treatments with various letters differ significantly ($p < 0.05$). Values in parenthesis are showing the percentage increase (+) or decrease (−) over blank treatment.

2.5. Effect of EC and AG Formulation of Strigolactone Analogs on Striga Infection under African Field Conditions

Finally, we investigated the efficacy of MP3 27EC and Nijmegen 34EC under naturally infested farmer-field conditions in INERA, Burkina Faso (Figure 8). In the pearl millet field, we observed 47–60% reduction in *Striga* emergence upon AG formulation of the two SL analogs application compared to blank treatment (Figure 8C). The EC formulation of Nijmegen 34EC showed about 25% reduction in *Striga* emergence in comparison to blank. Surprisingly, EC formulation of both SL analogs had a positive impact on pearl millet grain yield (126–137%). Moreover, MP3 in EC formulation showed a significant increase in grains per panicle over blank treatment (Supplementary Figure S1). In the sorghum farmer field, we observed 70–87% reduction in *Striga* emergence by SL analogs in AG formulation while EC formulation of MP3 and Nijmegen showed 35–65% reduction in *Striga* emergence (Figure 8E). This reduction in *Striga* emergence revealed a close association with *Striga* biomass, collected at final harvest from sorghum field (Supplementary Figure S2). We also observed 56–76% reduction in *Striga* biomass over blank treatment in AG formulated SL analogs while the reduction in EC formulated plots was 62–68% over blank treatment (Supplementary Figure S2). EC formulation treated plots of both SL analogs showed an increase in sorghum grain yield (164–216%) (Figure 8F). AG formulated MP3 showed considerable increase in yield components of sorghum compared to blank treatment (Supplementary Figure S3). Likewise, in ICRISAT, Niger, we observed a 91% reduction in *Striga* emergence by EC formulated MP3 27EC which was 27% higher than AG formulation of MP3 (66%) (Figure 9). Although MP3 27EC showed a better activity than Nijmegen 34EC, we observed a huge variation in *Striga* infestation among the plots of each treatment.

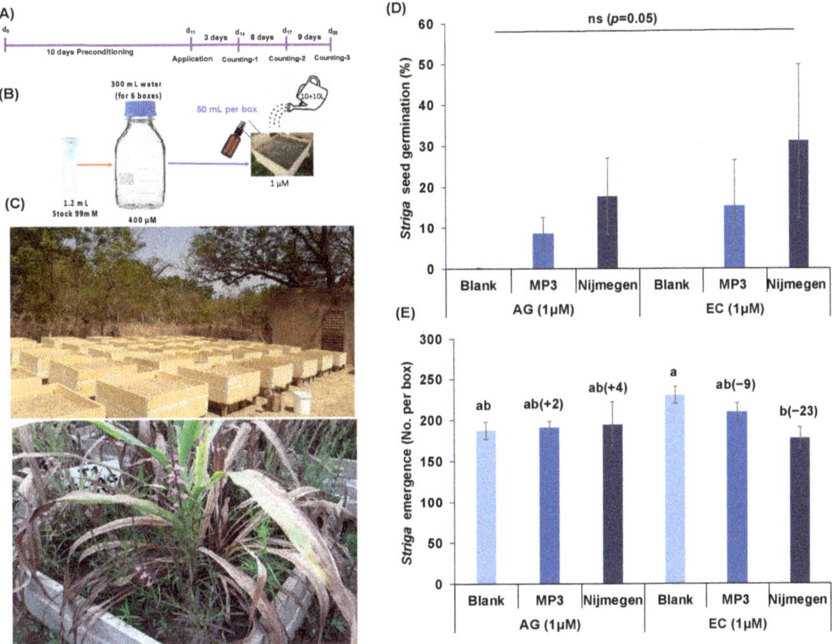

Figure 6. Effect of various formulations of SL analogs on *Striga* infection under mini-field conditions at Burkina Faso. (**A**) Scheme of the experiment conducted for *Striga* germination and emergence. (**B**) Experimental protocol adopted to conduct mini-field trials. Each box was filled with the soil, infested with *Striga* seeds collected from a pearl millet field in Burkina Faso. (**C**) View of the miniboxes and *Striga* infestation. (**D**) Average number of *Striga* seed germination in response to formulated MP3 and Nijmegen application. Both SL analogs of EC and AG formulations (at 1.0 µM) were applied in *Striga* infested mini-boxes. *Striga* germination was counted at 6 and 9 days after application. (**E**) Average number of *Striga* emergence in response to formulated MP3 and Nijmegen application. *Striga* emergence was counted after 80 days of pearl millet planting. Values of each bar showed average of *Striga* germination/emergence per minibox. Data are means $\pm$ SE ($n = 6$). For each SL analog, treatments with various letters differ significantly ($p < 0.05$). Values in parenthesis are showing the percentage increase (+) or decrease (−) over blank treatment. ns: non-significant.

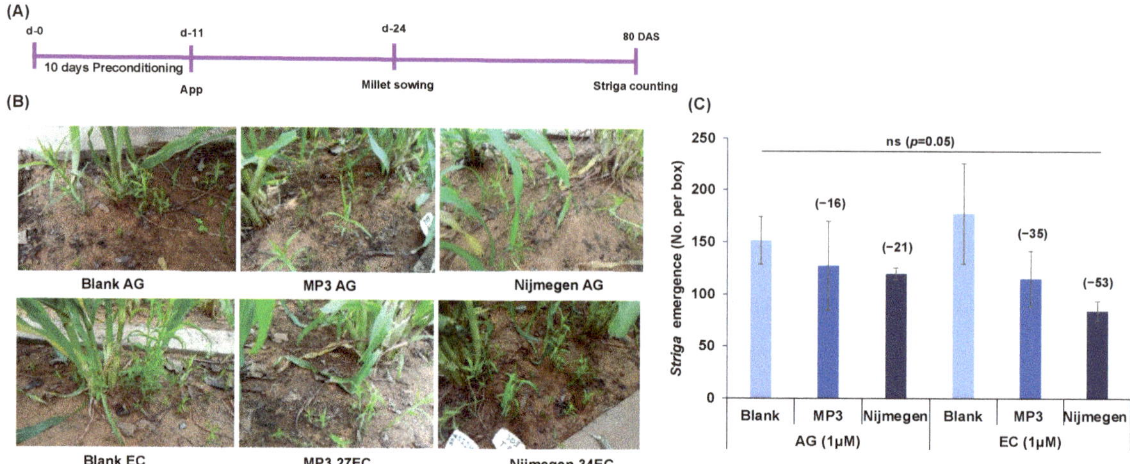

Figure 7. Effect of various formulations of SL analogs on *Striga* emergence under mini-field conditions at ICRISAT, Niger. (**A**) Scheme of the experiment conducted for *Striga* germination and emergence. (**B**) View of miniboxes and *Striga* infestation. Each box was filled with the soil, infested with *Striga* seeds collected from a pearl millet field in Niger. (**C**) Average number of *Striga* emergence in response to formulated MP3 and Nijmegen application. *Striga* emergence was counted after 80 days of pearl millet planting. Values of each bar showed average of *Striga* emergence per minibox. Data are means $\pm$ SE (n = 3). Values in parenthesis are showing the percentage increase (+) or decrease (−) over blank treatment. ns: non-significant.

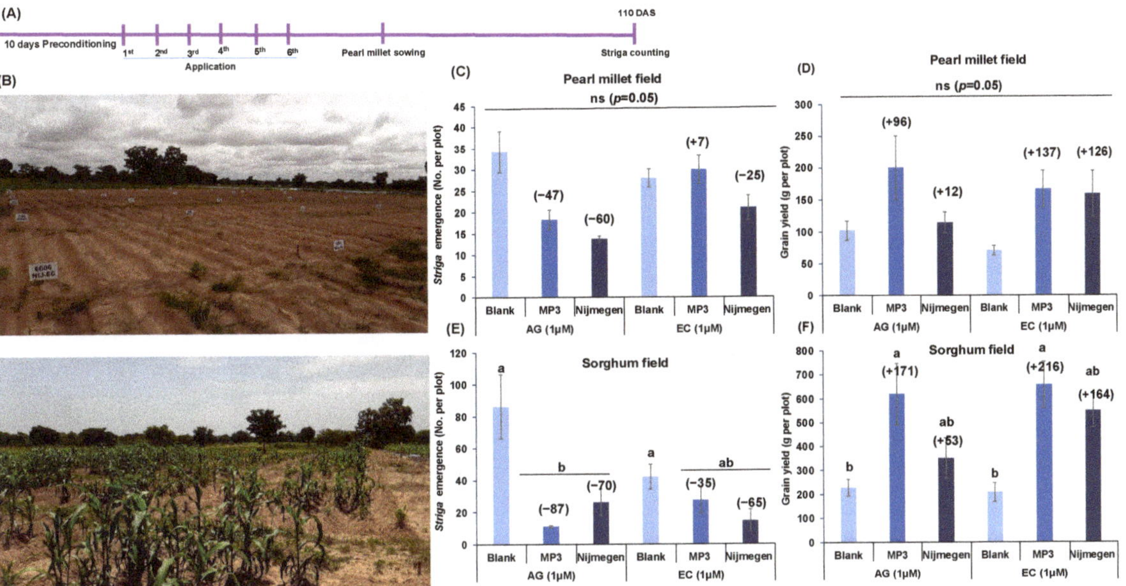

Figure 8. Effect of various formulations of SL analogs on *Striga* emergence and grain yield under naturally infested farmers field conditions at INERA, Burkina Faso. (**A**) Scheme of the experiment conducted for *Striga* emergence under farmers field conditions during 2020. (**B**) View of the *Striga* infested farmers field at Fada N'gourma, Burkina Faso. (**C**) Average number of *Striga* emergence in the pearl millet field in response to formulated MP3 and Nijmegen application. (**D**) Average grain yield per plot from the pearl millet field in response to formulated MP3 and Nijmegen application. (**E**) Average number of *Striga* emergence in the sorghum field in response to formulated MP3 and Nijmegen application. (**F**) Average grain yield per plot from the sorghum field in response to formulated MP3 and Nijmegen application. Both SL analogs of EC and AG formulations (at 1.0 µM) were applied in the *Striga* infested pearl millet and sorghum farmer fields. *Striga* emergence was counted after 110 days of pearl millet and sorghum planting. Values of each bar showed average of *Striga* emergence per plot. Data are means ± SE (*n* = 4). For each SL analog, treatments with various letters differ significantly (*p* < 0.05). Values in parenthesis are showing the percentage increase (+) or decrease (−) over blank treatment. ns: non-significant.

Figure 9. Effect of EC and AG formulations of the two SL analogs (MP3 and Nijmegen) on *Striga* emergence under artificially infested field conditions at ICRISAT, Niger. (**A**) Scheme of the experiment conducted for *Striga* emergence under field conditions. (**B**) View of the *Striga* infested farms under various treatments during 2020. (**C**) Average number of *Striga* emergence in the pearl millet field in response to formulated MP3 and Nijmegen application. (**D**) Average grain yield per plot from the pearl millet field in response to formulated MP3 and Nijmegen application. Both SL analogs of EC and AG formulations (at 1.0 µM) were applied in the *Striga* infested pearl millet field. *Striga* emergence was counted after 103 days of pearl millet planting. Values of each bar showed average of *Striga* emergence per plot. Data are means ± SE (*n* = 5). Values in parenthesis are showing the percentage increase (+) or decrease (−) over blank treatment. ns: non-significant.

3. Discussion

Striga, an obligate parasitic plant, attaches to the root system of most cereal crops in Africa while its subterranean nature of damage has made its control very difficult [5,35]. Developing suitable control strategies to minimize these underground losses has been proposed and advocated during the past few decades [28,36]. However, the underground damage of the host crop can be decreased only by reducing the seed bank density of *Striga* in the infested soil [37]. The "Suicidal Germination" technology has been suggested and tested in some field studies but with a lot of obstacles and limited success [38,39]. We have been working for the last five years to overcome the challenges of suicidal technology. We found two potent SL analogs MP3 and Nijmegen that can be applied into the field as suicidal agents [8,9]. In addition to the simple and efficient germination stimulants, the selection of suitable formulation is critical to facilitate large-scale field application [23]. A good formulation of SL analogs not only enhances the efficacy but also increases the stability and shelf-life after application into the field [22,40]. Atlas G-1086, a polyoxyethylene sorbitol hexaoleate mixed in cyclohexanone, has been used to formulate SL analogs in some past studies [23,25,26]. Previously, this emulsifier had been used to formulate Nijmegen-1 to apply in tobacco fields parasitized by *Orobanche ramosa* [22,41]. Moreover, some other potent SL analogs were also formulated with AG and used in *Striga* infested pearl millet and sorghum farmers field in Burkina Faso, showing a promising impact on reduction in *Striga* emergence [23,25]. However, there were few drawbacks of using AG to formulate SL analogs. We have been working closely with our partner, UPL, India, on developing suitable, handy, effective, and easily accessible formulated SL analogs that confer not only stability but also the ability to deplete easily of the seed bank in the infested soil. For this purpose, a new formulation "Emulsifiable Concentrate (EC)" of MP3 and Nijmegen (Figure 1) has been prepared by UPL, India.

In this report, we investigated the newly developed EC formulation of MP3 and Nijmegen with in vitro lab bioassays studies and the results indicated that MP3 and Nijmegen would be the potent SL analogs on seed germination (Figure 2). Both analogs with EC formulation showed better induction of *Striga* germination and EC_{50} over AG formulation. The induction of germination of various batches of *Striga* seeds by EC formulation of both SL analogs suggested that this formulation can be equally effective against all ecotypes of *Striga* in various parts of Africa (Figure 3). However, efficacy of SL analogs for both formulations varied for different *Striga* seed populations, with highest germination (60%) in Sudan batch and lowest germination (11%) in Niger batch. This variation among various populations might be attributed to seed viability, dormancy, as well as response to germination stimulants. The bioactivity of EC formulated MP3 for a longer period of time also indicates its potential as a good suicidal agent for field application in African soil. The EC formulated SL analogs possess longer shelf life and stability that is the desired characteristics for real field applications (Figure 4). In our previous studies, we observed several adverse effects of AG formulation on plant development and growth, which can be overcome by the EC formulation. Moreover, it is hard to dissolve SL analogs in AG formulation as compared to EC formulation. We also validated our lab outcome in a pot study under greenhouse conditions. Both SL analogs with EC and AG formulation have shown 85–99% decrease in *Striga* emergence (Figure 5).

However, we failed to get a significant impact on the reduction of *Striga* emergence under mini-field conditions both at INERA, Burkina Faso and ICRISAT, Niger (Figure 6). In fact, we had only one application in the mini-field, which might not have been enough to successfully reach to the *Striga* seeds to induce germination. In addition, a high variation among mini-fields might be due to several reasons, including high *Striga* density in the infested soils, low efficacy of applied treatments, soil type, leakage of water along the side of mini-fields, insufficient gap between application time, and host planting. It is recommended to enhance the number of applications in the mini box (>1) and to give enough time for suicidal death after *Striga* germination. Indeed, when we increased the number to six applications under naturally infested farmer fields in Burkina Faso, we observed very encouraging results, particularly in the sorghum field (Figure 8). Surprisingly, AG formulated SL analogs showed 70–87% reduction in *Striga* emergence in sorghum field and 47–60% in pearl millet field while EC formulated SL analogs did not reveal any significant impact on *Striga* emergence. A possible reason behind this low activity of EC formulated SL analogs is that the host crop was planted just after one week of the last application. Since EC formulated SL analogs are more stable so they may remain active in the soil for a longer period of time, which might continuously induce *Striga* germination that is easily attached to the host root. It is recommended that the host crop should not be grown just after the last application of EC-formulated SL analogs. We should consider enough time (3–4 months) or the whole rainy season for maximum induction of suicidal germination of *Striga* seeds. In spite of this, reduction in *Striga* infection led to better yield of both sorghum and pearl millet crops (Figure 8). In ICRISAT, Niger although it is an artificial infested field, we observed a huge variation among replicated plots (Figure 9). The emergence of *Striga* is limited such that was is hard to make conclusions from these findings. Low viability of *Striga* seeds, seed dormancy, cross-contamination among the plots, insect attack on the host, and poor growth of host plant can be possible reasons. We will repeat these field trials and expand our work in Kenya and Sudan to validate the findings, which will also expand our technology tackles different *Striga* ecotypes among Africa.

In summary, our results clearly demonstrate the effectiveness of the newly developed EC formulation of two potent and simple synthetic SL analogs as suicidal agents for practical field application. The new EC formulation of the two SL analogs appeared to be very bioactive, showing significant *Striga* reduction of 89–99% in rice under greenhouse conditions and 35–65% reduction in sorghum under field conditions in Burkina Faso. The reduction in *Striga* infection also led to increase rice plant height (61–71%) and pearl millet grain yield (>200%) as compared to blank treatment. The new EC formulation of the two SL analogs appeared to be very bioactive in terms of *Striga* reduction and host yield increase. In addition, the other advantages of EC formulation are simplicity, large-scale easy synthesis, stability, friendly use, easy packing/transportation and storage at normal room temperature. The development of proposed EC formulated simple SL analogs is the first step towards the large-scale synthesis of suicidal agents for field application and we believe this product will bring a breakthrough in suicidal technology to combat *Striga* in Africa.

4. Materials and Methods

4.1. Plant Materials and Chemicals

The SLs analogs MP3 and Nijmegen were synthesized and provided by Prof. Binne Zawanenburg, Radboud University, The Netherlands. Atlas G-1086 (provided by Croda Crop Care, Gouda, The Netherlands) was mixed with cyclo-hexanone (1:4) to make conventional formulation of SL analogs, used previously in the field. The new formulation "Emulsifiable Concentrate" (EC) of the two SL analogs (MP3, Nijmegen, The Netherlands) were prepared by UPL, India and Safety Data Sheet (SDS) has been attached (Supplementary Files S1 and S2). MP3 27EC means that 27 mg MP3 has been dissolved in 1.0 mL EC solvent (98.4 mM) while Nijmegen 34EC means that 34 mg Nijmegen has been added in 1.0 mL EC solvent (99 mM). *Striga hermonthica* seeds were collected from a sorghum (*Sorghum bicolor*) field during 2020 in Sudan (provided by Prof. A. G. Babiker), a maize (*Zea mays*) field during 2018 in Kenya (provided by Prof. Steven Runo, Kenyata University, Nairobi, Kenya), a pearl millet (*Pennisetum glaucum*) field during 2020 in Burkina Faso (provided by Dr. Djibril Yonli, INERA) and a pearl millet field during 2019 in Niger (provided by Dr. Mohammed Riyazaddin, ICRISAT, Niamey, Niger). Seeds of rice IAC-165 were obtained from Africa Rice, Tanzania (Courtesy of Dr. Jonne Rodenburg).

4.2. Striga Seed Germination Bioassays

First of all, the two formulations of both SL analogs were tested under lab conditions. *Striga* germination bioassays was conducted by following the procedure as described before [42]. The *Striga* seeds (collected from a sorghum, maize, and pearl millet infested field in Sudan, Kenya, Burkina Faso, and Niger, respectively) were first pre-conditioned before treating with SL analogs. For this purpose, the seeds were surface sterilized with 50% commercial bleach for 6–7 min and washed subsequently six times with MiliQ water in a laminar flow cabinet. The surface sterilized seeds (~50–100) were spread uniformly on a glass fiber filter paper disc (9 mm). Then 3 mL of sterilized MiliQ water was added in a plastic Petri plate containing one sterilized Whatman filter paper and 12 discs with *Striga* seeds. The Petri plates were sealed with parafilm, wrapped in aluminum foil, and incubated at 30 °C for 10 days. On the 11th day, the discs were dried in a laminar flow cabinet and the SL analogs (50 μL) were applied on each disc with various concentrations ranging from 10^{-5} M to 10^{-11} M. After application, the *Striga* seeds were induced to germinate in the dark for 24 h at 30 °C. The discs were scanned under a binocular microscope and germinated, and non-germinated seeds were counted by SeedQuant [43] and germination rate (in %) was calculated.

4.3. Stability Analysis

The stability of both the formulations of two selected SL analogs was observed in an in vitro bioassay. Five filter papers were added in plastic Petri plates (9 cm). Then, 10 mL of MP3 or Nijmegen with EC- and AG formulations at 1.0 μM concentration were added in each Petri plate and sealed with parafilm to incubate in the dark at 30 °C for two-week intervals for up to 12 weeks. On the final week, five glass fiber filter paper discs, containing 50–100 pre-conditioned *Striga* seeds were added in each plate and incubated at 30 °C for 24 h.

The discs were scanned under a binocular microscope and germinated and non-germinated seeds were counted by SeedQuant [43] and germination rate (in %) was calculated.

4.4. Striga Emergence in Pots under Greenhouse Conditions

The biological activity of the two formulations of the SL analogs was further tested in pots under greenhouse conditions as described [25,32,44]. A sand and soil (Stender, Basissubstrat) mixture (1:3 ratio) was prepared. About 0.5 L of this mixture without *Striga* seeds was added in the bottom of a 3 L perforated plastic pot. Then about 20 mg (Approximately 8000) *Striga* seeds (collected from a sorghum field during 2020 in Sudan) were thoroughly mixed in 1.5 L soil mixture and added on the top of clean soil in the same pot. The pots were given light irrigation under greenhouse conditions at 30 °C for pre-conditioning of *Striga* seeds for 10 days. Then each pot was treated with 500 mL (1.0 µM) of various treatments to allow *Striga* seeds to germinate without host for another 10 days. Then one week old three rice seedlings (IAC-165) were planted in the middle of each pot. The rice plants were allowed to grow under normal growth condition (30 °C, 65% RH). After 70 days of sowing, *Striga* emergence was observed in each pot and compared with mock treatment.

4.5. Striga Emergence under Mini-Box Conditions

The selected two SL analogs with both formulations were also evaluated in mini-boxes at INERA, Burkina Faso and at ICRISAT, Niger. At INERA, a 1 m × 1 m and 40 cm deep wooden box was used while in ICRISAT, Niger a mini-field, made of cemented bricks measuring 1 m × 2 m was used to assess *Striga* germination and emergence in response to SL treatments. The top 10 cm of soil in the box was infested with 50 mg of *Striga* seeds. Moreover, nine eplee bags containing surface sterilized *Striga* seeds (30–50 on average) were buried at a 10-cm depth in the box in an equidistant position and allowed them to pre-condition under hot and humid conditions for 10 days. At the end of the pre-conditioning of *Striga* seeds, both formulations of the SL analogs were applied (at 1.0 µM) and blank treatments (AG-1086 and EC) were included for comparison. Three eplee bags were taken out at 3, 6, and 9 days after SL application to count germinated and non-germinated seeds. After two weeks of application, pearl millet was sown in each box and emerged *Striga* plants were counted at 80 days after sowing (DAS).

4.6. Striga Emergence under Field Conditions

The two formulations of the candidate SL analogs were further assessed under naturally infested farmers field conditions in Eastern Burkina Faso and artificially infested field in ICRISAT, Niger. In Burkina Faso, two highly *Striga*-infested fields located near Kouaré research station (11°58′49″ N, 0°18′30″ E) of INERA were selected. In each field trial, the plots (4 × 4 m^2) with various treatments were laid out by following randomized complete block design (RCBD) with six replications. Each plot comprised of five ridges spaced 80-cm apart. All of the plots were spaced with four (4) blank ridges to avoid any contamination of the treatments. The two formulations (EC vs. AG) of MP3 and Nijmegen were applied (25 mL/m^2 at 400 µM) for six times, after onset of rainfall ($\geq$10 mm) to make the final concentration of 1.0 µM. We included blank treatment as a control to compare the treatment effects. In the blank, we added the same amount of EC or AG emulsifier without SL analogs (active ingredients). Pearl millet (local cultivar Idipiéni) and sorghum (local cultivar Itchoari) crops were sown at 2 weeks after the last application. *Striga* emergence was counted at 110 days after crop planting. In ICRISAT, Niger the field was prepared with ploughing and planking and ridges with 80-cm spaces were made. Each plot consisted of 4 ridges (4 × 4 m^2) and each ridge was infested with ~1.0 g *Striga* seeds. The trial was laid out by following RCBD with six replications. The field was artificially irrigated for pre-conditioning of *Striga* seeds for 10 days. All of the plots were treated with various treatments for six times and pearl millet (SOSTA-C88-P10) was sown at two weeks after the last application. *Striga* emergence was counted at 103 days after host planting.

4.7. Statistical Analysis

All of the data were collected by following standard procedure and collected data were analyzed statistically using statistical software package R (version 3.2.2). One-way analysis of variance (ANOVA) with Least Significant Difference (LSD) multiple range test and unpaired *t*-test were used for analyzing the effect of two formulations of the SL analogs on *Striga* infestation.

Supplementary Materials: The following supporting information can be downloaded at: https://www.mdpi.com/article/10.3390/plants11060808/s1, Supplementary Figure S1: Effect of various formulations of SL analogs on number of grains per panicle, panicle dry biomass, total stalk yield and harvest index in Pearl millet field at INERA, Burkina Faso. (**A**) Average grains per panicle (**B**) Average weight of dry panicle (**C**) Average stalk yield and (**D**) Average of harvest index from Striga infested Pearl millet field in response to formulated MP3 and Nijmegen application. Both SL analogs of EC and AG formulations (at 1.0 µM) were applied in Striga infested Pearl millet. Yield and yield component were measured after 110 days of pearl millet planting. Data are means ± SE (n = 5). For each SL analog, treatments with various letters differ significantly (p < 0.05).Values in parenthesis are showing the percentage increase (+) or decrease (-) over blank treatment. *ns*: non-significant. Figure S2: Effect of various formulations of SL analogs on Striga dry biomass at Burkina Faso. Average Striga dry biomass emerged in Sorghum field in response to formulated MP3 and Nijmegen application. Both SL analogs of EC and AG formulations (at 1.0 µM) were applied in Striga infested Sorghum fields. *Striga* seedlings were collected from Sorghum infested field and *Striga* dry biomass was measured at final harvest. Values of each bar showed average of Striga dry biomass emerged per plot. Data are means ± SE (n = 5). For each SL analog, treatments with various letters differ significantly (p < 0.05). Values in parenthesis are showing the percentage increase (+) or decrease (-) over blank treatment. ns: non-significant. Figure S3: Effect of various formulations of SL analogs on number of grains per panicle, panicle dry biomass, total stalk yield and harvest index from *Striga* infested Sorghum field at INERA, Burkina Faso. (**A**) Average grains per panicle, (**B**) Average weight of panicle, (**C**) Average stalk yield and (**D**) Average of harvest index from *Striga* infested Sorghum field in response to formulated MP3 and Nijmegen application. Both SL analogs of EC and AG formulations (at 1.0 µM) were applied in *Striga* infested Sorghum field. Yield and yield component were measured after 110 days of Sorghum planting. Values of each bar showed average of *Striga* emergence per plot. Data are means ± SE (n = 5). For each SL analog, treatments with various letters differ significantly (p < 0.05). Values in parenthesis are showing the percentage increase (+) or decrease (-) over blank treatment. *ns*: non-significant.

Author Contributions: S.A.-B. and M.J. conceived and designed the experiments. M.J., J.Y.W., L.B. and G.-T.E.C. performed lab and greenhouse studies. D.Y., O.M., H.T., M.R. and P.G. conducted mini-box and field experiments. R.H.P., S.E.B. and B.Z. helped in synthesis of SL analogs. J.Y.W., M.J. and S.A.-B. wrote the manuscript. All authors have read and agreed to the published version of the manuscript.

Funding: This study was supported by the Bill & Melinda Gates Foundation (grant number OPP1136424 to S.A.), and King Abdullah University of Science and Technology (KAUST), Saudi Arabia.

Informed Consent Statement: Not applicable.

Data Availability Statement: Not applicable.

Acknowledgments: We are grateful to Abdel Gabar Babiker, The National Research Center, Sudan; Steven Runo, Kenyatta University, Kenya for *Striga* seeds; We are thankful to Jonne Rodenburg, Africa Rice, Tanzania for providing seeds of rice IAC-165.

Conflicts of Interest: We are providing financial support to UPL company from our project funded by Bill & Melinda Gates Foundation. There is not any conflict of interest with UPL company.

References

1. Ejeta, G. Breeding for Striga Resistance in Sorghum: Exploitation of an Intricate Host–Parasite Biology. *Crop Sci.* **2007**, *47*, 216–229. [CrossRef]
2. Parker, C. Observations on the Current Status of *Orobanche* and *Striga* Problems Worldwide. *Pest Manag. Sci.* **2009**, *65*, 453–459. [CrossRef] [PubMed]
3. Spallek, T.; Mutuku, M.; Shirasu, K. The Genus Striga: A Witch Profile. *Mol. Plant Path.* **2013**, *14*, 861–869. [CrossRef]

4.   Jamil, M.; Kountche, B.A.; Al-Babili, S. Current Progress in Striga Management. *Plant Physiol.* **2021**, *185*, 1339–1352. [CrossRef] [PubMed]
5.   Scholes, J.D.; Press, M.C. Striga Infestation of Cereal Crops–an Unsolved Problem in Resource Limited Agriculture. *Curr. Opin. Plant Biol.* **2008**, *11*, 180–186. [CrossRef]
6.   Rodenburg, J.; Riches, C.R.; Kayeke, J.M. Addressing Current and Future Problems of Parasitic Weeds in Rice. *Crop Prot.* **2010**, *29*, 210–221. [CrossRef]
7.   Ejeta, G.; Gressel, J. *Integrating New Technologies for Striga Control: Towards Ending the Witch-Hunt*; World Scientific: Singapore, 2007; pp. 3–6.
8.   Atera, E.A.; Itoh, K.; Azuma, T.; Ishii, T. Farmers' Perspectives on the Biotic Constraint of Striga hermonthica and its Control in Western Kenya. *Weed Biol. Manag.* **2012**, *12*, 53–62. [CrossRef]
9.   Pennisi, E. Armed and Dangerous. *Science* **2010**, *327*, 804–805. [PubMed]
10.   Pennisi, E. How Crop-Killing Witchweed Senses its Victims. *Science* **2015**, *350*, 146–147. [CrossRef]
11.   Rubiales, D.; Verkleij, J.; Vurro, M.; Murdoch, A.J.; Joel, D. Parasitic Plant Management in Sustainable Agriculture. *Weed Res.* **2009**, *49*, 1–5. [CrossRef]
12.   Berner, D.; Kling, J.; Singh, B. Striga Research and Control. A Perspective from Africa. *Plant Dis.* **1995**, *79*, 652–660. [CrossRef]
13.   Hearne, S.J. Control—the Striga Conundrum. *Pest Manag. Sci.* **2009**, *65*, 603–614. [CrossRef]
14.   Bebawi, F.F.; Eplee, R.E.; Harris, C.E.; Norris, R.S. Longevity of Witchweed (*Striga asiatica*) Seed. *Weed Sci.* **1984**, *32*, 494–497. [CrossRef]
15.   Abayo, G.O.; English, T.; Eplee, R.E.; Kanampiu, F.K.; Ransom, J.K.; Gressel, J. Control of Parasitic Witchweeds (*Striga* spp.) on Corn (*Zea mays*) Resistant to Acetolactate Synthase Inhibitors. *Weed Sci.* **1998**, *46*, 459–466. [CrossRef]
16.   Matusova, R.; Rani, K.; Verstappen, F.W.; Franssen, M.C.; Beale, M.H.; Bouwmeester, H.J. The Strigolactone Germination Stimulants of the Plant-Parasitic *Striga* and *Orobanche* spp. are Derived from the Carotenoid Pathway. *Plant Physiol.* **2005**, *139*, 920–934. [CrossRef]
17.   Matusova, R.; van Mourik, T.; Bouwmeester, H.J. Changes in the Sensitivity of Parasitic Weed Seeds to Germination Stimulants. *Seed Sci. Res.* **2004**, *14*, 335–344. [CrossRef]
18.   Al-Babili, S.; Bouwmeester, H.J. Strigolactones, a Novel Carotenoid-Derived Plant Hormone. *Ann. Rev. Plant Biol.* **2015**, *66*, 161–186. [CrossRef]
19.   Xie, X.N.; Yoneyama, K.; Yoneyama, K. The Strigolactone Story. *Ann. Rev. Phytopath.* **2010**, *48*, 93–117. [CrossRef]
20.   Fiorilli, V.; Wang, J.Y.; Bonfante, P.; Lanfranco, L.; Al-Babili, S. Apocarotenoids: Old and New Mediators of the Arbuscular Mycorrhizal Symbiosis. *Front. Plant Sci.* **2019**, *10*, 1186–1195. [CrossRef]
21.   Sibhatu, B. Review on Striga Weed Management. *Int. J. Life Sci. Sci. Res.* **2016**, *2*, 110–120.
22.   Zwanenburg, B.; Mwakaboko, A.S.; Kannan, C. Suicidal Germination for Parasitic Weed Control. *Pest Manag. Sci.* **2016**, *72*, 2016–2025. [CrossRef] [PubMed]
23.   Kountche, B.A.; Jamil, M.; Yonli, D.; Nikiema, M.P.; Blanco-Ania, D.; Asami, T.; Zwanenburg, B.; Al-Babili, S. Suicidal Germination as a Control Strategy for *Striga hermonthica* (Benth.) in Smallholder Farms of Sub-Saharan Africa. *Plants People Planet* **2019**, *1*, 107–118. [CrossRef]
24.   Uraguchi, D.; Kuwata, K.; Hijikata, Y.; Yamaguchi, R.; Imaizumi, H.; Sathiyanarayanan, A.; Rakers, C.; Mori, N.; Akiyama, K.; Irle, S. A Femtomolar-Range Suicide Germination Stimulant for the Parasitic Plant *Striga hermonthica*. *Science* **2018**, *362*, 1301–1305. [CrossRef] [PubMed]
25.   Jamil, M.; Kountche, B.A.; Wang, J.Y.; Haider, I.; Jia, K.-P.; Takahashi, I.; Ota, T.; Asami, T.; Al-Babili, S. A New Series of Carlactonoic Acid Based Strigolactone Analogs for Fundamental and Applied Research. *Front. Plant Sci.* **2020**, *11*, 1–13. [CrossRef] [PubMed]
26.   Samejima, H.; Babiker, A.G.; Takikawa, H.; Sasaki, M.; Sugimoto, Y. Practicality of the Suicidal Germination Approach for Controlling Striga hermonthica. *Pest Manag. Sci.* **2016**, *72*, 2035–2042. [CrossRef] [PubMed]
27.   Mwakha, F.A.; Budambula, N.L.; Neondo, J.O.; Gichimu, B.M.; Odari, E.O.; Kamau, P.K.; Odero, C.; Kibet, W.; Runo, S. Witchweed's Suicidal Germination: Can Slenderleaf Help? *Agronomy* **2020**, *10*, 873. [CrossRef]
28.   Parker, C.; Riches, C. *Parasitic Weeds of the World: Biology and Control*; CAB International: Wallingford, UK, 1993; pp. 1–74.
29.   Fukui, K.; Ito, S.; Asami, T. Selective Mimics of Strigolactone Actions and Their Potential Use for Controlling Damage Caused by Root Parasitic Weeds. *Mol. Plant* **2013**, *6*, 88–99. [CrossRef] [PubMed]
30.   Blanco-Ania, D.; Mateman, J.J.; Hylova, A.; Spichal, L.; Debie, L.M.; Zwanenburg, B. Hybrid-type Strigolactone Analogues Derived from Auxins. *Pest Manag. Sci.* **2019**, *75*, 3113–3121. [CrossRef]
31.   Jamil, M.; Kountche, B.A.; Haider, I.; Wang, J.Y.; Aldossary, F.; Zarban, R.A.; Jia, K.-P.; Yonli, D.; Hameed, U.F.S.; Takahashi, I.; et al. Methylation at the C-3' in D-Ring of Strigolactone Analogs Reduces Biological Activity in Root Parasitic Plants and Rice. *Front. Plant Sci.* **2019**, *10*, 1–14. [CrossRef]
32.   Jamil, M.; Kountche, B.A.; Haider, I.; Guo, X.J.; Ntui, V.O.; Jia, K.-P.; Ali, S.; Hameed, U.F.S.; Nakamura, H.; Lyu, Y.; et al. Methyl Phenlactonoates are Efficient Strigolactone Analogs with Simple Structure. *J. Exp. Bot.* **2018**, *69*, 2319–2331. [CrossRef]
33.   Jia, K.-P.; Kountche, B.A.; Jamil, M.; Guo, X.; Ntui, V.O.; Rüfenacht, A.; Rochange, S.; Al-Babili, S. Nitro-phenlactone, a Carlactone Analog with Pleiotropic Strigolactone Activities. *Mol. Plant* **2016**, *9*, 1341–1344. [CrossRef] [PubMed]

34. Zwanenburg, B.; Blanco-Ania, D. Strigolactones: New Plant Hormones in the Spotlight. *J. Exp. Bot.* **2018**, *69*, 2205–2218. [CrossRef] [PubMed]

35. Jamil, M.; Charnikhova, T.; Verstappen, F.; Bouwmeester, H. Carotenoid Inhibitors Reduce Strigolactone Production and *Striga hermonthica* Infection in Rice. *Arch. Biochem. Biophys.* **2010**, *504*, 123–131. [CrossRef] [PubMed]

36. Kim, S.; Adetimirin, V.; The, C.; Dossou, R. Yield Losses in Maize Due to *Striga hermonthica* in West and Central Africa. *Int. J. Pest Manag.* **2002**, *48*, 211–217. [CrossRef]

37. Oswald, A. Striga Control—Technologies and Their Dissemination. *Crop Prot.* **2005**, *24*, 333–342. [CrossRef]

38. Wigchert, S.; Kuiper, E.; Boelhouwer, G.; Nefkens, G.; Verkleij, J.; Zwanenburg, B. Dose−Response of Seeds of the Parasitic Weeds *Striga* and *Orobanche* toward the Synthetic Germination Stimulants GR24 and Nijmegen 1. *J. Agric. Food Chem.* **1999**, *47*, 1705–1710. [CrossRef]

39. Yoneyama, K.; Awad, A.A.; Xie, X.; Yoneyama, K.; Takeuchi, Y. Strigolactones as Germination Stimulants for Root Parasitic Plants. *Plant Cell Physiol.* **2010**, *51*, 1095–1103. [CrossRef]

40. Kgosi, R.L.; Zwanenburg, B.; Mwakaboko, A.S.; Murdoch, A.J. Strigolactone Analogues Induce Suicidal Seed Germination of *Striga* spp. in Soil. *Weed Res.* **2012**, *52*, 197–203. [CrossRef]

41. Zwanenburg, B.; Mwakaboko, A.S.; Reizelman, A.; Anilkumar, G.; Sethumadhavan, D. Structure and Function of Natural and Synthetic Signalling Molecules in Parasitic Weed Germination. *Pest Manag. Sci.* **2009**, *65*, 478–491. [CrossRef] [PubMed]

42. Jamil, M.; Kanampiu, F.K.; Karaya, H.; Charnikhova, T.; Bouwmeester, H.J. *Striga hermonthica* Parasitism in Maize in Response to N and P Fertilisers. *Field Crops Res.* **2012**, *134*, 1–10. [CrossRef]

43. Braguy, J.; Ramazanova, M.; Giancola, S.; Jamil, M.; Kountche, B.A.; Zarban, R.A.Y.; Felemban, A.; Wang, J.Y.; Lin, P.-Y.; Haider, I.; et al. SeedQuant: A Deep Learning-Based Tool for Assessing Stimulant and Inhibitor Activity on Root Parasitic Seeds. *Plant Physiol.* **2021**, *186*, 1632–1644. [CrossRef] [PubMed]

44. Wang, J.Y.; Jamil, M.; Lin, P.-Y.; Ota, T.; Fiorilli, V.; Novero, M.; Zarban, R.A.; Kountche, B.A.; Takahashi, I.; Martínez, C. Efficient Mimics for Elucidating Zaxinone Biology and Promoting Agricultural Applications. *Mol. Plant* **2020**, *13*, 1654–1661. [CrossRef] [PubMed]

Article

Good News for Cabbageheads: Controlling *Phelipanche aegyptiaca* Infestation under Hydroponic and Field Conditions

Amit Wallach [1,2,*], Guy Achdari [1] and Hanan Eizenberg [1]

[1] Department of Phytopathology and Weed Research, Agricultural Research Organization, Newe Ya'ar Research Center, Ramat Yishay 3009503, Israel; achdarig@volcani.agri.gov.il (G.A.); eizenber@volcani.agri.gov.il (H.E.)

[2] The Robert H. Smith Institute of Plant Sciences and Genetics, The Robert H. Smith Faculty of Agriculture, Food and Environment, The Hebrew University of Jerusalem, Rehovot 7610001, Israel

[*] Correspondence: aa.wallach@gmail.com

Abstract: *Phelipanche aegyptiaca* (Orobanchaceae) is a parasitic weed that causes severe yield losses in field crops around the world. After establishing vascular connections to the host plant roots, *P. aegyptiaca* becomes a major sink that draws nutrients, minerals, and water from the host, resulting in extensive crop damage. One of the most effective ways to manage *P. aegyptiaca* infestations is through the use of herbicides. Our main objective was to optimize the dose and application protocol of herbicides that effectively control *P. aegyptiaca* but do not damage the cabbage crop. The interactions between the cabbage roots and the parasite were first examined in a hydroponic system to investigate the effect of herbicides on initial parasitism stages, e.g., germination, attachment, and tubercles production. Thereafter, the efficacy of glyphosate and ethametsulfuron-methyl in controlling *P. aegyptiaca* was examined in five cabbage fields naturally infested with *P. aegyptiaca*. The herbicides glyphosate and ethametsulfuron-methyl were applied on cabbage foliage and in the soil solution, both before and after the parasite had attached to the host roots. A hormesis effect was observed when glyphosate was applied at a dose of 36 g ae ha^{-1} in a non-infested *P. aegyptiaca* field. Three sequential herbicide applications (21, 35, and 49 days after planting) effectively controlled *P. aegyptiaca* without damaging the cabbages at a dose of 72 g ae ha^{-1} for glyphosate and at all the examined doses for ethametsulfuron-methyl. Parasite control with ethametsulfuron-methyl was also effective when overhead irrigation was applied after the herbicide application.

Keywords: *Phelipanche aegyptiaca*; glyphosate; ethametsulfuron-methyl; chemical control

Citation: Wallach, A.; Achdari, G.; Eizenberg, H. Good News for Cabbageheads: Controlling *Phelipanche aegyptiaca* Infestation under Hydroponic and Field Conditions. *Plants* **2022**, *11*, 1107. https://doi.org/10.3390/plants11091107

Academic Editors: Fabrizio Araniti, Yaakov Goldwasser and Evgenia Dor

Received: 14 January 2022
Accepted: 15 April 2022
Published: 19 April 2022

Publisher's Note: MDPI stays neutral with regard to jurisdictional claims in published maps and institutional affiliations.

1. Introduction

Egyptian broomrape *Phelipanche aegyptiaca* (Orobanchaceae) is a root parasitic plant that has a wide range of hosts in field crops. As such, it is considered a major troublesome weed that causes severe yield losses around the world [1]. *P. aegyptiaca* is very common in the Mediterranean region and East Asia, and there are also reports that it has taken hold in parts of Australia, Europe, and the Americas [2]. Ambient conditions in Israel, with its Mediterranean climate, support the development of *P. aegyptiaca* in vegetable fields [3]. For all the broomrapes (*Phelipanche* and *Orobanche* spp.), including *P. aegyptiaca*, the seeds can lie dormant in the soil for a long time, even decades, and then germinate after pre-conditioning in response to germination stimulants [4,5]. After seed germination, the parasitic plant invades the host's root vascular system, and once established, the parasite draws all its nutritional and water requirements from the host, resulting in extensive damage to the host plant.

One of the most effective ways to manage *P. aegyptiaca* infestations in the field is via the application of herbicides. However, the herbicide-based control of *P. aegyptiaca* in the field is a complex task, for two main reasons: the herbicide must be selective, i.e., not damage the host plants (since it will move through the host to the parasitic plant), and the types of herbicides that can be applied to the parasitic broomrapes are limited to particular target sites, e.g., photosystem inhibitors are not effective because of the absence of PSII in the parasitic plants [5]. Herbicides based on one of two modes of action are thus used to control *P. aegyptiaca*, i.e., through the inhibition of aromatic amino acid biosynthesis (by targeting 5-enolpyruvylshikimate-3-phosphate synthase (EPSPS); group 9 HRAC), e.g., glyphosate, which is generally applied to the foliage, or through the inhibition of branched-chain amino acid biosynthesis (by targeting acetolactate synthase (AHAS/ALS); group 2 HARC), e.g., imidazolinones, sulfonylureas, and pyrithiobac-sodium, which are usually applied to the soil for absorption by the roots. Therefore, the site of herbicide application must be chosen according to its mode of action [5].

In the proposed study, in which we investigated herbicides suitable for application in *P. aegyptiaca*-infested cabbage fields, we were in a position to leverage the knowledge acquired over the years by the Eizenberg group in the management of broomrape infestations of a variety of crop species [5], as reviewed in brief below. Goldwasser et al. (2001) studied the control of *P. aegyptiaca* parasitism in potato fields by the sequential application of the ALS inhibitors imazapic and rimsulfuron. They found that three sequential applications of imazapic on potato foliage 20, 40, and 60 days after full potato emergence effectively controlled *P. ramosa* but impaired the quality of the potatoes. By contrast, the application of rimsulfuron followed by overhead irrigation controlled *P. aegyptiaca* efficiently and did not harm the potatoes. The difference in response to the two herbicides may derive from the availability of rimsulfuron in the rhizosphere, which allows the direct translocation of the herbicide to the *P. aegyptiaca* attachments and may, therefore, prevent crop damage [6].

Cochavi et al. [7] developed a protocol to control *P. aegyptiaca* in carrot fields using foliar herbicide applications. They found that glyphosate and imazapic in doses lower than 108 g ae ha^{-1} and 2.4 g ai ha^{-1}, respectively, did not harm the carrot taproot biomass in non-infested fields at low doses of herbicide. They also found that the yields of the carrots classified as class A were higher for sequential applications of glyphosate vs. imazapic. By contrast, imazamox at all examined doses led to a reduction in carrot yield. Therefore, they decided to continue the experiment with only glyphosate applied to the foliage by monitoring *P. aegyptiaca* development using a minirhizotron system when *P. aegyptiaca* attachments were 2 mm, and sequential applications were within 21 days intervals. When the carrot biomass was determined, a hormesis response was observed at a glyphosate concentration of 137 g ae ha^{-1} in a *P. aegyptiaca*-infested carrot field. However, three sequential applications of glyphosate at low doses of 54 or 108 g ae ha^{-1} were successful in controlling *P. aegyptiaca* [7].

Aly et al. (2001) conducted a study aiming to control *O. cumana* (sunflower broomrape) in the field. They found that sunflower vigor was impaired by two sequential applications of imazapic on sunflower foliage 12 cm and 55 cm tall with treatments at doses of 3.6 or 4.8 g ai ha^{-1}. However, when they used drip irrigation and reduced the imazapic dose, the herbicide still gave effective control, facilitating an increase in the yield [8]. Eizenberg [9] also cooperated with a group in Oregon, USA [10], in a study aiming to control small broomrape (*O. minor*) in red clover fields. Using a growing degree days (GDD) model, they examined the effect of imazamox at 800 and 1000 GDD and found that all the examined doses controlled *O. minor* effectively, with the largest attachments being better controlled than the small attachments.

Some 20 years later, Eizenberg and Goldwasser [11] developed a holistic decision support system (DSS), which they designated 'PICKIT', to control *P. aegyptiaca* in fields of processing tomatoes [12]. The *P. aegyptiaca* management tools of the DSS are based on parasitism dynamics models [12,13] that estimate key parasitism stages in terms of GDD, foliar herbicide applications, and drip chemigation and then provide the optimal timing, doses, and methodology of herbicide application. Finally, Cohen et al. used GIS to approximate and evaluate the *P. aegyptiaca* infestation level in the field and on a regional scale [14].

Building on the knowledge acquired by the Eizenberg group, the objectives of the current study were twofold: (i) to investigate the effect of herbicides belonging to two different classes of compounds, glyphosate and ethametsulfuron-methyl (EMS), on cabbage yield; (ii) to optimize the dosage and application practice of the herbicides to effectively control *P. aegyptiaca* in cabbage fields. Both herbicides are considered appropriate for use in cabbage.

2. Materials and Methods

2.1. Plant Material

2.1.1. Cabbage

The white cabbage (*Brassica oleracea*) cultivars 'Froctor' (Zeraim Gedera-Syngenta, Israel) 'Fresco', and 'Cheers' (Eden Seeds, Moshav Hatzav, Israel) and the red cabbage cultivar 'Grand-Rio' (Tarsis Agrichem, Petah Tikvah, Israel) were used in this study. All cabbage plants that were used in the experiments were planted out 30 days after seeding in trays.

2.1.2. Phelipanche Aegyptiaca

P. aegyptiaca inflorescences were collected in 2017 from a cabbage field in southern Israel. After the seeds had been sieved through a 300-micron mesh, they were stored at 4 °C in the dark until use.

2.2. Laboratory and Field Experiments

2.2.1. Herbicides

Glyphosate (Roundup^{TM}, 360 g ae /L) was obtained from Bayer, Monsanto Company (St. Louis, MO, USA), and ethametsulfuron-methyl (Salsa^{TM} 75% WG) from Du Pont Inc. (Wilmington, DE, USA). EMS was mixed with alkylaryl polyether alcohol (DX, 800 g/L; Adama-Agan, Ashdod, Israel).

2.2.2. Germination Test

P. aegyptiaca seeds were surface-sterilized for 3 min in 70% ethanol and then for 10 min in 1% sodium hypochlorite before washing with distilled water [15]. The seeds were kept in a sterile hood chamber until dry and then stored at 4 °C in the dark until use. For the germination experiment, the seeds were spread on 8 mm Whatman$^{®}$ glass microfiber filter disks, Grade GF/A (Whatman International, Maidstone, UK), held in Petri dishes. The disks were wetted with 31 μL of distilled water, and the Petri dishes were sealed with Parafilm strips. The experiment was conducted in a chamber held in the dark at 25 °C. After 7 days of preconditioning, 28 μL of GR$_{24}$ (StrigoLab, Turin, Italy) at a concentration of 10^{-6} μL/mL was added to each disk; demineralized water was used as the control. Four days after GR$_{24}$ was added, *P. aegyptiaca* germination was determined with a binocular electronic microscope (Leica M80, Wetzlar, Germany) [16].

2.2.3. Experiments in Polyethylene Bags

In addition to the above experiments, experiments in polyethylene bags (PEB) were performed to investigate *P. aegyptiaca* herbicide response at the subsurface stages, both pre- and post-attachment to the host's roots, according to the method described by Parker and Dixon [17]. Briefly, cabbage (cultivar 'Froctor') seedlings were re-rooted in 250 mL of 5% Hoagland's solution [18]. When the seedlings had developed new root systems, the plants were placed on 35 × 24 mm GF/A glass microfiber paper sheets, and *P. aegyptiaca* seeds were spread uniformly over the paper sheets [17]. Thereafter, 1 mL of herbicide at a concentration of 5×10^{-3} mL/L for glyphosate or 0.125 g/L for EMS was applied using a manual sprinkler over the cabbage foliage immediately after the parasite was attached to the host. Three and six days after herbicide application, healthy attachments were counted and classified according to Perez-de-Luque et al. (2016) [19]. Control efficacy is presented as a percentage of total healthy attachments of the non-treated control.

When herbicide was applied in the pre-attachment parasitism stages, 1.5 mL of glyphosate or EMS solution was applied as described above or by syringing 5 mL of EMS onto the GF/A paper sheet. When the herbicide was applied in the pre-attachment parasitism experiment, *P. aegyptiaca* seeds were counted seven days after herbicide application, and *P. aegyptiaca* establishment was assessed as a percent of the necrotic attachments out of the total number of attachments.

2.2.4. Field Experiments

Seven field experiments were conducted in commercial cabbage fields, some naturally infested with *P. aegyptiaca* and others not infested, at different locations in Israel—Sde Tzvi (experiments designated A, B, and D), Nahalal (experiment C), Beit Hagedi (experiments E and F), and Mevo Hama (experiment G). The experiments were conducted in blocks in a random factorial design. Each plot was 5 m long and 1.93 m wide. The herbicides were applied sequentially 21, 35, and 49 days after planting (DAP) at 200 L ha^{-1} with a motorized GKS15 sprayer equipped with a Tee Jet$^®$ 110015 nozzle (Maruyama Mfg CO. Inc., Chiyoda-ku, Japan) and operated at a pressure of 300 kPa. In experiments A and B, each herbicide was tested at six different doses for foliar application as follows: glyphosate 36, 72, 144, 188, 432, and 576 g ae ha^{-1} and EMS 4, 9, 18, 28, 37, and 75 g ai ha^{-1}. The control plots were not treated with herbicide. In experiments C, D, E, and F, glyphosate and EMS were tested for a foliar application on the basis of the doses that were found in experiments A and B, namely, 72 g ae ha^{-1} and 18.5 g ai ha^{-1}, respectively. In experiment G, 300 m^3 ha^{-1} water was applied with overhead irrigation after each herbicide application at 21, 35, and 49 DAP within eight hours after herbicide application. At the conclusion of the field experiments, cabbage heads were harvested from an area of 4 m^2 in each plot.

2.3. Statistical Analyses

All data sets were analyzed using RStudio Version 1.4.1717 (RStudio team). ANOVA was performed if the data showed a normal distribution. The normality of the data was determined using the Shapiro–Wilk test, and the means were compared using Tukey's honestly significant difference (HSD) ($p < 0.05$). Data sets with non-normal distributions were analyzed using a general linear model (GLM), and least-squares means were computed into the package "EMMEANS" [20]. Means in the PEB experiment three and six days after herbicide application were compared using a paired *t*-test.

Dose–response (hormesis) was incorporated into the "drc" [21] in R as follows:

$$y = c + \frac{(d - c + fx)}{1 + exp(b(log(x) - log(e)))} \tag{1}$$

b and *e* have no direct interpretation (while *b* and *e* are constants), *c* represents the lower horizontal asymptote, *d* represents the upper horizontal asymptote, and *f* is the size of the hormesis effect. The resulting model is a four-parameter log-logistic model [22].

3. Results

3.1. Effect of Herbicides on P. aegyptiaca in Polyethylene Bags

3.1.1. Herbicide Application Post-Attachment

Effective control (vs. untreated cabbage plants) was obtained by foliar spraying of *P. aegyptiaca*-infected cabbages (growing under hydroponic conditions) with glyphosate or EMS when the broomrape tubercles reached 2.5 mm in size (Figure 1). For the herbicide-treated plants, there were 89% and 91% healthy *P. aegyptiaca* attachments for glyphosate and EMS, respectively, three days after the treatment, compared with 27% and 29% healthy attachments, respectively, six days after the treatment (Figure 1).

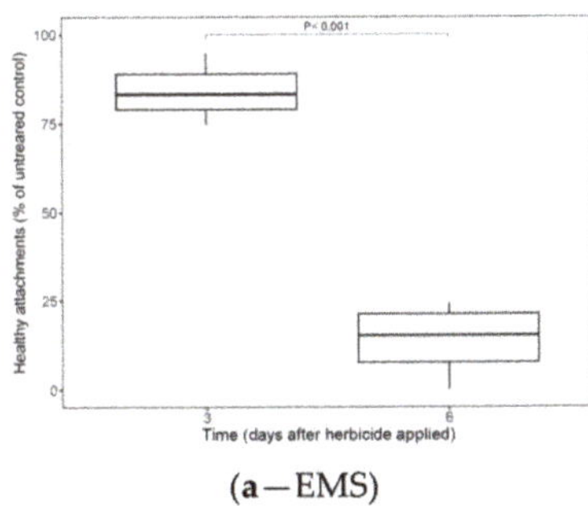
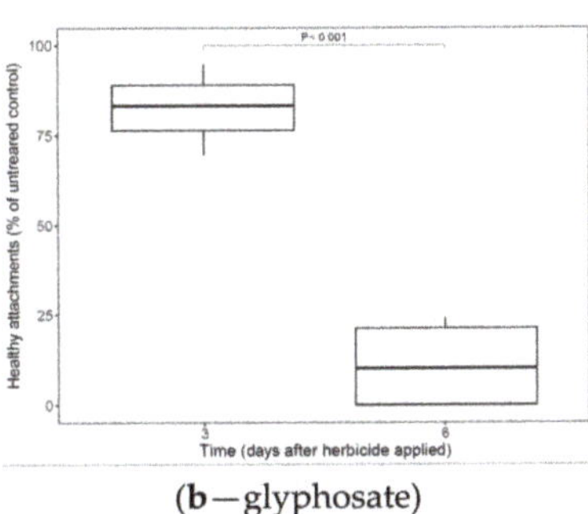

(a — EMS)   **(b — glyphosate)**

Figure 1. Healthy attachments of *P. aegyptiaca* compared with non-treated control in cabbage grown hydroponically in polyethylene bags. Ethametsulfuron-methyl (EMS, **a**) and glyphosate (**b**) were applied to the cabbage foliage after *P. aegyptiaca* attachment. Means were compared using a paired *t*-test ($p < 0.05$); the numbers in the figure represent *p*-values.

3.1.2. Herbicide Applications Pre-Attachment

When herbicides were applied in the pre-parasitism stage, both glyphosate and EMS markedly impaired *P. aegyptiaca* attachment (Figure 2). Excellent control of *P. aegyptiaca* was achieved by the injection of EMS into the root zone, namely, 100% of the seeds showed necrosis. For the foliar application of glyphosate and EMS, 82% and 75%, respectively, of seeds showed necrosis, compared with 10% necrotic seeds in the non-treated control (Figure 2).

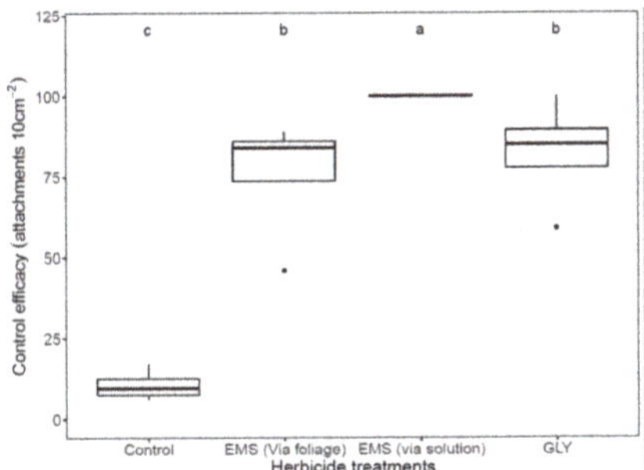

Figure 2. Control efficacy of *P. aegyptiaca* in cabbage grown hydroponically in polyethylene bags by applying the herbicide, glyphosate (GLY), or ethametsulfuron-methyl (EMS), before *P. aegyptiaca* attachment to the cabbage roots (vs. the non-treated control). Means were compared using GLM. ($p < 0.05$); means labeled with the same letter do not show significant differences.

3.2. Field Experiments

The seven experiments that were conducted under field conditions aimed to determine the optimal herbicide doses that would not damage the cabbages (in the fields not infested with *P. aegyptiaca*) and optimize *P. aegyptiaca* control efficacy in cabbage fields naturally infested with *P. aegyptiaca*.

3.3. Cabbage Response to Herbicide under Field Conditions

3.3.1. Fitted Herbicide Dose for Cabbage Safety

Experiments A and B were performed in Sde Tzvi to investigate the dose–response effect of glyphosate and EMS on the yield of the host plant. For glyphosate, the sensitivity of the cabbages to the herbicide was reflected by a hormesis effect, which describes the relationship between low doses of glyphosate and cabbage ('Froctor') yield (Figure 3a).

Three sequential foliar applications to the cabbage at a dose of 72 g ae ha^{-1} glyphosate on 21, 35, and 49 DAP resulted in a slight reduction in cabbage yield (91% compared with the non-treated control). When the hormesis effect was computed at a lower dose of 36 g ae ha^{-1} of glyphosate, the yield increased to 134% compared with the untreated control. For glyphosate doses >72 g ae ha^{-1}, the cabbage yield decreased markedly—to as little as 13% compared with the non-treated control. Likewise, cabbage development was inhibited at higher glyphosate doses (Figure 3a). Cabbage yield was not reduced (vs. the non-treated control) for any of the examined sequential applications of EMS (Figure 3b).

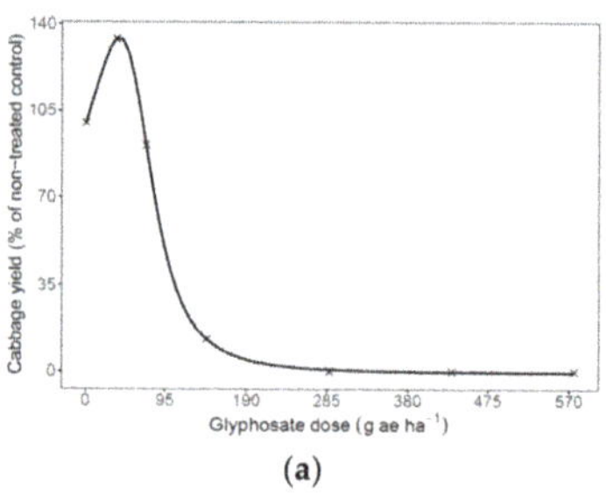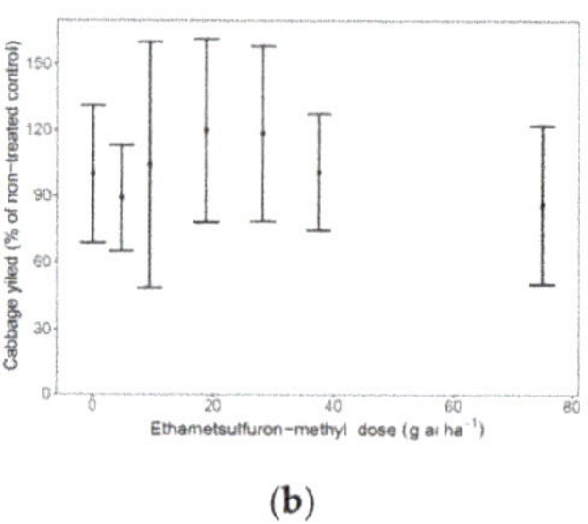

(**a**)　　　　　　　　(**b**)

Figure 3. Sensitivity of cabbages to the herbicides glyphosate (**a**) and ethametsulfuron-methyl (EMS, **b**) applied sequentially on 21, 35, and 49 DAP. (**a**) A four-parameter modified sigmoid equation (for u-shaped hormesis) was fitted to the cabbage response to glyphosate; the parameters were as follows: upper asymptote (d = 100, SE = 0.74, $p \leq 0.0001$), 50% of cabbage yield (X_0 = 94.83, SE = 0.06, $p \leq 0.0001$), and size of the hormesis effect (f = 11.63, SE = 0.38, $p \leq 0.0001$). (**b**) Relationship between cabbage yield and EMS dose. Since there were no significant differences in the response of the cabbages to the various doses of EMS, a regression equation was not fitted.

3.3.2. Herbicide Applications for Cabbage Save

There was no reduction in cabbage ('Froctor') yield in experiment C (Nahalal) when glyphosate or EMS was sequentially applied to the cabbage leaves at the doses of 72 g ae ha^{-1} glyphosate and 18 g ai ha^{-1} EMS; cabbage yields in this field were 103% and 105%, respectively, compared with non-treated control (Figure 4).

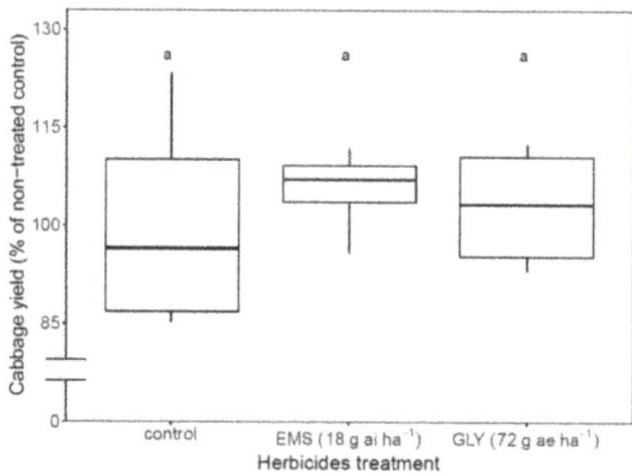

Figure 4. Cabbage sensitivity to the herbicides glyphosate (GLY) and ethametsulfuron-methyl (EMS). Herbicides were applied 21, 35, and 49 DAP at the doses of 72 g ae ha^{-1} glyphosate and 18 g ai ha^{-1} EMS. Means were compared using the Tukey HSD test ($p < 0.05$); means that are indicated with the same letter do not show significant differences.

3.4. Herbicide-Based Control of P. aegyptiaca under Field Conditions

To optimize the herbicide dose for *P. aegyptiaca* control in cabbage fields naturally infested with *P. aegyptiaca*, in experiments D, E, and F (Sde Tzvi and Beit Hagedi), glyphosate and EMS were applied at doses of 72 g ae ha^{-1} and 18 g ai ha^{-1}, respectively, on the three white cabbage cultivars. When glyphosate was applied to the cabbage foliage, no *P. aegyptiaca* shoots were observed in the field, and full *P. aegyptiaca* control was achieved for all three cultivars (Figure 5a–c). However, EMS was less successful in controlling the parasite, with 114 ('Froctor'), 58 ('Fresco'), and 104 ('Cheers') *P. aegyptiaca* shoots appearing per 4 m^2, similar to the non-treated control blocks (Figure 5).

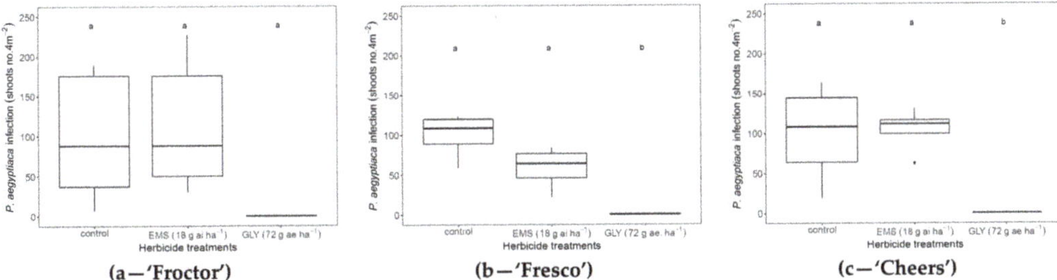

(a – 'Froctor')   (b – 'Fresco')   (c – 'Cheers')

Figure 5. Control efficacy of *P. aegyptiaca* in three cabbage cultivars, 'Froctor' (**a**), 'Fresco' (**b**), and 'Cheers' (**c**). The herbicides glyphosate (GLY) and ethametsulfuron-methyl (EMS) were applied sequentially at doses of 72 g ae ha^{-1} and 18 g ai ha^{-1}, respectively, to the cabbage foliage at 21, 35, 49 DAP. Means were compared using the Tukey HSD test ($p < 0.05$); means that are indicated with the same letter do not show significant differences. The experiment was performed in cabbage fields naturally infested with *P. aegyptiaca* at Sde Tzvi (**a**) and Beit Hagedi (**b**,**c**).

Herbicide Adjustment Method Using Overhead Irrigation System

In experiment G (Mevo Hama), foliar application of the herbicide at doses of 72 g ae ha^{-1} and 18 g ai ha^{-1} for glyphosate and EMS, respectively, was followed by sprinkler irrigation in an amount of 300 m^3 ha^{-1}; the herbicide was thus incorporated into the soil and taken up through the leaves and roots. In fields naturally infested with *P. aegyptiaca*, the herbicides were applied to two cabbage varieties, 'Froctor' and 'Grand-Rio'. No *P. aegyptiaca* shoots appeared when EMS was applied to both varieties. However, when glyphosate was applied, the average numbers of *P. aegyptiaca* shoots counted were 1.754 m^{-2} and 0.254 m^{-2} *P. aegyptiaca* shoots for 'Froctor' and 'Grand-Rio' cultivars, respectively, compared with 104 m^{-2} and 324 m^{-2} for the non-treated plot, respectively (Figure 6).

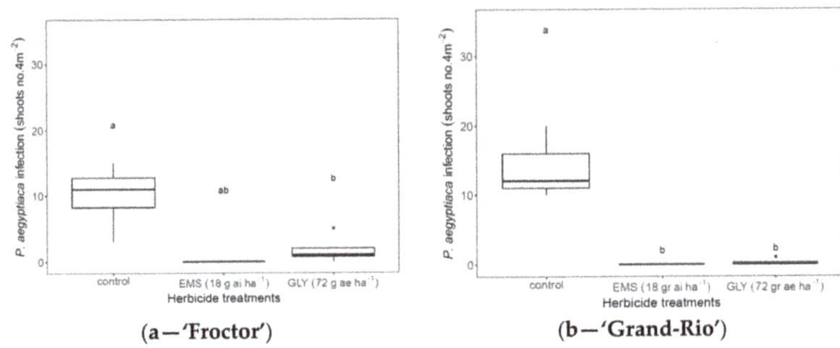

(a – 'Froctor')   (b – 'Grand-Rio')

Figure 6. In *P. aegyptiaca*-infested fields, cabbage cultivars 'Froctor' (**a**) and 'Grand-Rio' (**b**) were treated with three sequential foliar applications of ethametsulfuron-methyl (EMS) or glyphosate (GLY) 21, 35, and 49 DAP at the doses of 72 g ae ha^{-1} glyphosate and 18 g ai ha^{-1} EMS, followed by water spraying after each herbicide eight hours after herbicide application. Means were compared using GLM ($p < 0.05$); means that are indicated with the same letter do not show significant differences.

4. Discussion

The herbicide-based management of *Orobanche* and *Phelipanche* broomrape species in the field is a complicated task because of the unique life cycle and biology of these root parasitic plants. In particular, herbicides that target PSII are not suitable because of the fact that the parasite is lacking a photosynthetic system. Thus, the types of herbicides that can be used are those whose modes of action inhibit the biosynthesis of aromatic (e.g., glyphosate) or branched-chain (e.g., EMS) amino acids [5]. Furthermore, controlling *Phelipanche* in the field must be carried out at the initial soil-subsurface developmental stages, e.g., attachments, tubercle development, and tubercles with crown roots (spider stage) [12,19,23]. Applying herbicides after *Phelipanche* shoots emerge is too late to prevent crop damage, but it is nevertheless effective in preventing seed dispersal because of the sterilizing effect of the herbicide on the broomrape inflorescences [9,11]. There are thus two strategies for herbicide control of *P. aegyptiaca* in the field. The first is to apply the herbicide at the pre-attachment stage directly to the soil so as to prevent attachments or control the small tubercles in the soil sub-surface at the parasitism stage. The second is to apply a systemic herbicide that is translocated via the phloem after the parasite has attached to the host. The latter strategy is based on the functioning of *Phelipanche* species as powerful sinks that draw nutrients and water from the host plant [5].

The EPSPS inhibitor glyphosate is thus usually applied—according to the second strategy—at low doses to the host's foliage; from there, it moves through the host's vascular system to the strong broomrape sink [24]. Glyphosate has thus been used for controlling *Phelipanche* infestations in carrot and parsley fields [7,25]. In the current study, when glyphosate was applied to a cabbage field not infested with *P. aegyptiaca*, the cabbage yield was not damaged at doses lower than 72 g ae ha^{-1}, with the hormesis effect being detected at 36 g ae ha^{-1}. The hormesis response to glyphosate has been shown in other crops, for example, carrot, corn, soy, and barley [7,26,27]. Similar to the reported studies, our results exhibited effective *P. aegyptiaca* control in cabbage when low doses of glyphosate were applied in three sequential applications.

ALS enzyme inhibitors, such as imidazolinones and sulfonylureas, also effectively control species of *Phelipanche*. For example, the imidazolinone herbicides imazapic and imazamox have been used successfully for the control of *P. aegyptiaca* and *O. crenata* in parsley [24], *P. aegyptiaca* in tomato [5], *O. cumana* in sunflower [8,23], and *O. minor* in red clover [9]. Among the sulfonylureas, sulfosulfuron is licensed in Israel only for the control of *P. aegyptiaca* in processing tomatoes; it was thus used in a unique DSS for the rational management of the broomrape [11]. Another sulfonylurea, rimsulfuron, has been reported to effectively control *P. aegyptiaca* in potatoes and low infestations in fields of processing tomatoes [6,28]. In this study, yet another member of this group, EMS, was tested for the first time for the control of *P. aegyptiaca* in cabbage. Cabbage sensitivity to EMS was observed at doses higher than the recommended dose (18 g ai ha^{-1}).

Under hydroponic conditions (PEB system) in both application methods (foliar application and through the root solution), full control of *P. aegyptiaca* both pre- and post-attachment was observed. In the field, when EMS was applied on cabbage foliage without overhead irrigation (i.e., sprinklers), there was no control of *P. aegyptiaca*. Similar to the current study, Eizenberg et al. reported that sulfonylurea herbicides control *P. aegyptiaca* only when applied to the soil solution [28]. Two reasons may explain the conflicting results obtained with two different application methodologies for EMS. First, in the field, the herbicide was metabolized or excluded, and therefore, it did not reach *P. aegyptiaca*. By contrast, in the PEB system, the cabbage plants remained small, and the herbicide was able to reach the tubercles. Another explanation could be that in the PEB system, the cabbage exudes the herbicide from the roots to the rhizosphere, and the parasite is thus controlled through the soil solution. A similar hypothesis was proposed for the red clover–imazamox association, for which the herbicide was exuded from the roots of red clover to the rhizosphere to control the small broomrape [10].

Applying sulfonylurea herbicides to the soil solution to control the root parasitic plant in the soil sub-surface parasitism stage requires precise knowledge about the parasitism dynamics. Models that predict the parasitism dynamics have been proposed for *P. aegyptiaca* in tomatoes [23], *O. minor* in red clover [29], *O. cumana* in [23], *P. aegyptiaca* in carrot [30], and *O. crenata* in legumes [19]. Thus, the next step in our study of the control of *P. aegyptiaca* in cabbage is the development of the relevant model. In the meantime, the current study has indeed shown that applying EMS via an overhead irrigation system (to incorporate the herbicide into the rhizosphere) prevents the establishment of the parasitic weed in the host root system and guards against yield losses. Moreover, EMS applied at an herbicidal dose also controlled troublesome non-parasitic weeds, whereas a low dose of glyphosate effectively controlled *P. aegyptiaca* alone.

In summary, *P. aegyptiaca* could be controlled using sequential treatments of low glyphosate doses when applied on cabbage foliage and using herbicidal EMS doses when applied on cabbage foliage and incorporated into the soil.

Author Contributions: H.E. conceived the idea and supervised the project. A.W., G.A. and H.E. designed the experiments. A.W. and G.A. conducted experiments. A.W. analyzed data and wrote the manuscript. All authors have read and agreed to the published version of the manuscript.

Funding: This research was funded by the Chief Scientist of the Israel Ministry of Agriculture (Grant No. 200260).

Institutional Review Board Statement: Not applicable.

Informed Consent Statement: Not applicable.

Data Availability Statement: Not applicable.

Conflicts of Interest: The authors declare no conflict of interest. The funders had no role in the design of the study; in the collection, analyses, or interpretation of data; in the writing of the manuscript; or in the decision to publish the results.

References

1. Gressel, J.; Joel, D.M. Weedy Orobanchaceae: The problem. In *Parasitic Orobanchaceae: Parasitic Mechanisms and Control Strategies*; Joel, D.M., Gressel, J., Musselman, L.J., Eds.; Springer: Berlin/Heidelberg, Germany, 2013; Volume 9783642381, pp. 309–312.
2. Parker, C. Parasitic Weeds: A World Challenge. *Weed Sci.* **2012**, *60*, 269–276. [CrossRef]
3. Joel, D.M. Direct infection of potato tubers by the root parasite *Orobanche aegyptiaca*. *Weed Res.* **2007**, *47*, 276–279. [CrossRef]
4. López-Granados, F.; García-Torres, L. Longevity of crenate broomrape (*Orobanche crenata*) seed under soil and laboratory conditions. *Weed Sci.* **1999**, *47*, 161–166. [CrossRef]
5. Eizenberg, H.; Hershenhorn, J.; Ephrath, J.H.; Kanampiu, F. Chemical control. In *Parasitic Orobanchaceae: Parasitic Mechanisms and Control Strategies*; Joel, D.M., Gressel, J., Musselman, L.J., Eds.; Springer: Berlin/Heidelberg, Germany, 2013; pp. 415–432.
6. Goldwasser, Y.; Eizenberg, H.; Hershenhorn, J.; Plakhine, D.; Blumenfeld, T.; Buxbaum, H.; Golan, S.; Kleifeld, Y. Control of *Orobanche aegyptiaca* and *O. ramosa* in potato. *Crop Prot.* **2001**, *20*, 403–410. [CrossRef]
7. Cochavi, A.; Achdari, G.; Smirnov, E.; Rubin, B.; Eizenberg, H. Egyptian Broomrape (*Phelipanche aegyptiaca*) Management in Carrot under Field Conditions. *Weed Technol.* **2015**, *29*, 519–528. [CrossRef]
8. Aly, R.; Goldwasser, Y.; Eizenberg, H.; Hershenhorn, J.; Golan, S.; Kleifeld, Y. Broomrape (*Orobanche cumana*) Control in Sunflower (*Helianthus annuus*) with Imazapic. *Weed Technol.* **2001**, *15*, 306–309. [CrossRef]
9. Eizenberg, H.; Colquhoun, J.B.; Mallory-Smith, C.A. Imazamox application timing for small broomrape (*Orobanche minor*) control in red clover. *Weed Sci.* **2006**, *54*, 923–927. [CrossRef]
10. Colquhoun, J.B.; Eizenberg, H.; Mallory-Smith, C.A. Herbicide Placement Site Affects Small Broomrape (*Orobanche minor*) Control in Red Clover. *Weed Technol.* **2006**, *20*, 356–360. [CrossRef]
11. Eizenberg, H.; Goldwasser, Y. Control of egyptian broomrape in processing tomato: A summary of 20 years of research and successful implementation. *Plant. Dis.* **2018**, *102*, 1477–1488. [CrossRef]
12. Eizenberg, H.; Aly, R.; Cohen, Y. Technologies for Smart Chemical Control of Broomrape (*Orobanche* spp. and *Phelipanche* spp.). *Weed Sci.* **2012**, *60*, 316–323. [CrossRef]
13. Ephrath, J.E.; Hershenhorn, J.; Achdari, G.; Bringer, S.; Eizenberg, H. Use of Logistic Equation for Detection of the Initial Parasitism Phase of Egyptian Broomrape (*Phelipanche aegyptiaca*) in Tomato. *Weed Sci.* **2012**, *60*, 57–63. [CrossRef]
14. Cohen, Y.; Roei, I.; Blank, L.; Goldshtein, E.; Eizenberg, H. Spatial spread of the root parasitic weed *Phelipanche aegyptiaca* in processing tomatoes by using ecoinformatics and spatial analysis. *Front. Plant Sci.* **2017**, *8*, 973. [CrossRef] [PubMed]
15. Plakhine, D.; Joel, D.M. Ecophysiological consideration of *Orobanche cumana* germination. *Helia* **2010**, *33*, 13–18. [CrossRef]

16. Plakhine, D.; Ziadna, H.; Joel, D.M. Is seed conditioning essential for *Orobanche* germination? *Pest Manag. Sci.* **2009**, *65*, 492–496. [CrossRef]

17. Parker, C.; Dixon, N. The use of polyethylene bags in the culture and study of *Striga* spp. and other organisms on crop roots. *Ann. Appl. Biol.* **1983**, *103*, 485–488. [CrossRef]

18. Hoagland, D.R.; Arnon, D.I. *The Water-Culture Method for Growing Plants Without Soil: California Agricultural Experiment Station*; University of California, The College of Agriculture: Berkeley, CA, USA, 1950; Volume 347.

19. Pérez-De-Luque, A.; Flores, F.; Rubiales, D. Differences in crenate broomrape parasitism dynamics on three legume crops using a thermal time model. *Front. Plant Sci.* **2016**, *7*, 1910. [CrossRef]

20. Lenth, R.V. *Emmeans: Estimated marginal means, aka least-squares means. R Packag, version 1.6.1*; R Fondation for Statistical Computing: Vienna, Austria, 2021.

21. Ritz, C.; Strebig, J.C.; Ritz, M.C. *Package 'drc'*; Creative Commons: Mountain View, CA, USA, 2016.

22. Brain, P.; Cousens, R. An equation to describe dose responses where there is stimulation of growth at low doses. *Weed Res.* **1989**, *29*, 93–96. [CrossRef]

23. Eizenberg, H.; Hershenhorn, J.; Achdari, G.; Ephrath, J.E. A thermal time model for predicting parasitism of *Orobanche cumana* in irrigated sunflower—Field validation. *F. Crop. Res.* **2012**, *137*, 49–55. [CrossRef]

24. Shilo, T.; Zygier, L.; Rubin, B.; Wolf, S.; Eizenberg, H. The mechanism of glyphosate control of *Phelipanche aegyptiaca*. *Planta* **2016**, *244*, 1095–1107. [CrossRef]

25. Goldwasser, Y.; Eizenberg, H.; Golan, S.; Kleifeld, Y. Control of *Orobanche crenata* and *Orobanche aegyptiaca* in parsley. *Crop Prot.* **2003**, *22*, 295–305. [CrossRef]

26. Belz, R.G.; Cedergreen, N.; Sørensen, H. Hormesis in mixtures—Can it be predicted? *Sci. Total Environ.* **2008**, *404*, 77–87. [CrossRef] [PubMed]

27. Belz, R.G.; Piepho, H.P. Predicting biphasic responses in binary mixtures: Pelargonic acid versus glyphosate. *Chemosphere* **2017**, *178*, 88–98. [CrossRef] [PubMed]

28. Eizenberg, H.; Goldwasser, Y.; Golan, S.; Plakhine, D.; Hershenhorn, J. Egyptian Broomrape (*Orobanche aegyptiaca*) Control in Tomato with Sulfonylurea Herbicides—Greenhouse Studies. *Weed Technol.* **2004**, *18*, 490–496. [CrossRef]

29. Eizenberg, H.; Colquhoun, J.; Mallory-Smith, C. A predictive degree-days model for small broomrape (*Orobanche minor*) parasitism in red clover in Oregon. *Weed Sci.* **2005**, *53*, 37–40. [CrossRef]

30. Cochavi, A.; Rubin, B.; Achdari, G.; Eizenberg, H. Thermal time model for egyptian broomrape (*Phelipanche aegyptiaca*) parasitism dynamics in carrot (*daucus carota* L.): Field validation. *Front. Plant Sci.* **2016**, *7*, 1807. [CrossRef]

MDPI
St. Alban-Anlage 66
4052 Basel
Switzerland
Tel. +41 61 683 77 34
Fax +41 61 302 89 18
www.mdpi.com

Plants Editorial Office
E-mail: plants@mdpi.com
www.mdpi.com/journal/plants